Non-Imaging Microwave and Millimetre-Wave Sensors for Concealed Object Detection

Non-Imaging Microwave and Millimetre-Wave Sensors for Concealed Object Detection

Boris Y. Kapilevich
Ariel University, Ariel, Israel
Stuart W. Harmer
Manchester Metropolitan University, Manchester, United Kingdom
Nicholas J. Bowring
Manchester Metropolitan University, Manchester, United Kingdom

CRC Press
Taylor & Francis Group
Boca Raton London New York

CRC Press is an imprint of the
Taylor & Francis Group, an **informa** business

CRC Press
Taylor & Francis Group
6000 Broken Sound Parkway NW, Suite 300
Boca Raton, FL 33487-2742

First issued in paperback 2017

No claim to original U.S. Government works
Version Date: 20140730

ISBN 13: 978-1-138-07274-9 (pbk)
ISBN 13: 978-1-4665-7714-5 (hbk)

Visit the Taylor & Francis Web site at
http://www.taylorandfrancis.com

and the CRC Press Web site at
http://www.crcpress.com

Contents

Preface

The tragic events of September 11th and the growing global threat of terrorist attacks have driven research and development of effective countermeasures. According to analysis performed by the Homeland Security Research Corporation (HSRC), Washington, DC, "... new and maturing sensor and ICT technologies create new market niches and fresh business opportunities. These shifts lead a $305 billion 2011 market to a $546 billion by 2022." HSRC analysts have forecasted that similar growth is likely for the sector of homeland security focused on explosives trace detection (ETD).

Modern people-screening technologies and explosives trace detection methods are based on a number of fundamental physical and chemical principles, including:

- Ion mobility spectroscopy (IMS)
- Chemiluminescence (CL)
- Electron capture detectors (ECDs)
- Surface acoustic wave (SAW)
- X-ray backscattering
- Active microwave and millimetre-wave (mm-wave) imaging
- Passive microwave and mm-wave imaging
- Terahertz imaging
- Metal detection
- Liquid explosives screening
- Biometric people screening
- Metal oxide semiconductor (MOS)–based electrochemical sensing
- Molecularly imprinted polymer sensing

That such a variety of detecting methods and principles have been investigated clearly demonstrates that developing effective anti-terror countermeasures is a very difficult task. At present, there is no universal solution to this problem, nor is there likely to be such a solution in the foreseeable future. The current approach is to concentrate efforts on improving overall performance of existing detecting systems in terms of higher detection probability and lower false alarm rate, resulting in a slow but steady improvement in these technologies. Over the last decade the authors were involved in the design and testing of various microwave and mm-wave devices and systems aimed at improving the efficiency of detecting weapons and explosives concealed upon the human body.

One motivation of this book is to systemize the results obtained by the research teams of the MM-Wave Laboratory of Ariel University (Israel) and the Center for Sensing and Imaging in the School of Electrical Engineering at Metropolitan University of Manchester (UK). We believe that such a systematic view will help specialists involved in the design of detection or screening systems to improve upon these devices. Another important reason for writing this book comes from the understanding that mm-wave imaging devices suffer from high complexity and very high cost, which prove to be serious limitations to the wider uptake of these scanners. In typical screening scenarios, an exact knowledge of the shape of a concealed object is not an essential requirement, so preliminary screening of a human body using a non-imaging remote sensor may be sufficient for making a decision about the presence (or absence) of concealments. The screening process can then be accomplished less expensively in comparison with the application of millimetre-wave imaging systems. Also, non-imaging detectors can be realized as handheld, portable devices that can be employed without fixed-position detection portals, giving far greater flexibility of use. It should also be pointed out that non-imaging sensors don't have the so-called 'naked vision' effect typical for mm-wave imagers and don't violate human privacy.

Our discussion focuses on non-imaging mm-wave and microwave sensors applied to remote detection of both metallic and dielectric objects concealed on a human body in indoor and outdoor environments for homeland security applications.

The book consists of seven chapters. The introduction (Chapter 1) describes in general the concept of active and passive mm-wave sensing. Advantages and drawbacks of both approaches are considered too.

Chapter 2 provides a brief background of theory and phenomenology for mm-wave sensors and discusses these considerations for both passive and active systems.

Chapter 3 describes some configurations of active mm-wave sensors.

Chapter 4 presents the basic realizations of mm-wave active sensors operating in frequency-modulated continuous wave (FMCW) mode. Various configurations of such sensors are considered and compared. Examples of sensors in 94 and 330 GHz are given.

Chapter 5 describes the application of complex natural resonances to the detection and identification of concealed objects with microwave frequency radiation.

Chapter 6 examines different versions of passive mm-wave sensors based on heterodyne and direct power receivers. A detection algorithm estimating unimodality of the recorded search tracing has been suggested and tested. It allows us to reach detection probabilities up to 95% and more. A portable version of the passive sensor is described for indoor application at a distance of up to 3 m.

Chapter 7 focuses on the role of shielding effects inherent to typical common materials such as textiles, leathers, cartons, etc. Additional attenuation and interference caused by these materials in screening for concealed objects may lead to degrading the overall performance of the sensors, both active and passive. The knowledge of complex permittivities allows us to predict such effects. However, Rayleigh scattering must also be taken into consideration since the surface roughness parameter is comparable with wavelength. The modification of a free space method of permittivity determination is described, and an angular invariance criterion is formulated too.

We address the text to MS and PhD students, microwave and mm-wave engineers, as well as specialists involved in the solution of a variety of problems associated with numerous homeland security applications. Hopefully, readers will find in this book some learning examples and stimulus for their own initiatives in this exciting field.

Boris Y. Kapilevich
Ariel University, MM-Wave Laboratory, Israel

Stuart W. Harmer
Manchester Metropolitan University, School of Electrical Engineering, UK

Nicholas J. Bowring
Manchester Metropolitan University, School of Electrical Engineering, UK

Acknowledgements

Professor Boris Kapilevich would like to thank the following colleagues for their assistance and contributions in the process of preparing this book: Professor Yosef Pinhasi, Dean of Faculty of Engineering, Ariel University, for providing favorable research environments and active support for many studies performed in the MM-Wave Laboratory; Danny Hardon, Director of Homeland Securty Center, Ariel University, for coordinating efforts; Professor A. Gover, Tel Aviv University; and Dr. Moshe Einat, member of the Electrical and Electronics Engineering Department of Ariel University, for valuable discussions.

The following engineers and students made valuable contributions to Chapters 4, 6 and 7: B. Litvak, A. Shulzinger, O. Shotman, B. Kostjakovsky, M. Anisimov, R. Arusi and A. Etinger.

Dr. Stuart Harmer and Professor Nicholas Bowring acknowledge the contribution made to Chapters 3 and 5 by members of the Centre for Sensing and Imaging research group in the School of Science and Engineering at Manchester Metropolitan University, UK. In particular, they recognize the significant work carried out by Dr. David Andrews, Dr. Nacer Rezgui, Dr. Matthew Southgate, Dr. Sarah Smith and Dr. Shawn Cole. Without these people, it would not have been possible to design, build and test various systems that are detailed in this book.

About the Authors

Boris Y. Kapilevich was appointed head of the Applied Electromagnetic Department at Siberia State University of Telecommunications and professor of microwave engineering in 1988 (Novosibirsk, Russia). In 2003, he joined the Department of Electrical and Electronics Engineering at Ariel University (Israel) to lead the research activity in the area of millimeter-wave (mm-wave) imaging and detection conducted by the RF/Microwave/MM-Wave Laboratory, where he focused on the solution of homeland security problems. He earned his PhD in 1970 from Novosibirsk State Technical University (Russia) and Dr.Sc. (Tech.) in 1988 from Moscow Power Engineering Institute (Russia), both in the area of microwave engineering.. Dr. Kapilevich has published four books, dozens of technical articles in peer-reviewed editions and has made numerous presentations at national and international professional forums, many in the field of mm-wave technology. He holds 15 patents in the area of microwaves and mm-wave devices. The "Concealed Weapons Detector," patented by Dr. Kapilevich and coauthors, won the TechConnect World Innovation and Expo 2014 awards in Washington, DC. He is a senior member of Institute of Electical and Electronics Engineers.

Stuart W. Harmer holds a BSc in physics and astrophysics and a PhD in physics from the University of Sussex (UK). He is a fellow of the Institute of Physics. From 2000 to 2006, Dr. Harmer held the position of research fellow in photonics at Sussex University and later the position of senior research fellow at Queen Mary University of London. He has worked in the industry as a scientist for SELEX ES and was appointed reader in physics and engineering science at Manchester Metropolitan University in 2012. Dr. Harmer has authored and coauthored more than 50 publications and is an inventor on 20 patents.

Nicholas J. Bowring holds an MSc in experimental physics and a PhD in atomic and molecular physics, earned from the University of Manchester in 1985 and 1991, respectively. In 2007 he was made full professor of electronic engineering at Manchester Metropolitan University and is currently head of the Centre for Sensing and Imaging. Professor Bowring has published numerous journal papers and conference proceedings, is a named inventor on more than 20 patents, and has worked on a diverse range of projects, including millimetre and microwave remote sensing radar and image processing in the visible band for train rail inspection.

1

Introduction

The millimetre-wave (mm-wave) band extends from 30 to 300 GHz. The electromagnetic properties of clothing textiles (leathers, carton, etc.) at these frequencies are such that millimetre-wave radiation can pass through these materials without high attenuation [1]. At shorter wavelengths, for example, infrared radiation, attenuation is too great for penetration through clothing. In addition, because the wavelength of millimetre-wave radiation (10 mm–1 mm) is not too large, a highly directive antenna may be formed without unacceptably large aperture dimensions, allowing sensors to be realized which give reasonable spatial resolution yet are able to obtain information about objects concealed upon the human body, under typical clothing. The only other regions of the electromagnetic spectrum where these conditions of spatial resolution and low attenuation through clothing fabrics are met are at very high frequencies (i.e. x-ray) and in the submillimetre, or terahertz, band. In the x-ray region, the obvious safety issues associated with exposing people to ionizing radiation prevent deployment; recent use of low-dose backscatter x-ray machines for security screening of people in airports has caused widespread controversy [2], and these devices have been withdrawn from service. Such devices must also be active, since the thermal grey body emission of x-rays from objects at temperatures of ~300 K is far too low for passive sensors to be practicable. There are no safety concerns with passive systems, as they simply detect the thermal emission from the body and that from the surroundings which reflect from the body. In the submillimetre-wave (THz) region of the electromagnetic spectrum the radiation does not have sufficient photon energy to ionize atoms or molecules and so is considered far safer than x-rays. Clothing materials also have favourable transparency in the submillimetre-wave band, although attenuation is greater than is found in the millimetre-wavelength region and attenuation increases significantly with increasing frequency [3]. Passive and active sensors can be built in the submillimetre-wavelength region, but these devices are currently not as mature as their millimetre-wave counterparts, and they require expensive and often experimental components.

Millimetre-wave screening is currently realized, almost exclusively, by sensors which form images. Both active and passive millimetre-wave imaging systems are commercially available for security screening applications [4]. Imaging systems are a natural product, as human operators are able to more easily interpret imagery than other information, such as temporal or frequency domain waveforms. However, unlike in the visible and infrared wavelength regions of the electromagnetic spectrum, formation of focal plane arrays of receivers is currently prohibitively expensive and technologically challenging, although alternative technologies may well allow the realization of focal plane arrays in the near future [5]. Most commercially available imaging systems use several receiver antennae, each with its low noise amplifier and heterodyne mixer. Imaging is accomplished by spatially sampling the field of view using mechanically movable reflective optics to direct millimetre radiation emanating from different parts of the scene onto an array of multiple receivers. Such configurations are expensive to build, as the components are costly to produce. Active systems based on aperture synthesis are a common sight in modern airport security lanes [6].

Millimetre-wave imaging is a very well understood science and has obvious applications: concealed object detection, especially in the security space where people can be screened, at standoff ranges, for objects concealed under clothing; and for aircraft landing systems, where dust/sand or smoke may prevent pilots from visualizing a landing site. The science of millimetre-wave imaging is based on spatially mapping the distribution of radiometric contrast in a scene. For passive imaging, this contrast is a function of the temperature and emissivity distribution of the scene and the reflectance distribution of the scene. However, in an outdoor environment, where the vastly different apparent temperature of the sky (very cold relative to room temperature) illuminates the scene and contributes to the contrast in the image formed, there is variability in image contrast depending on prevailing climatic conditions. Active millimetre-wave imaging uses artificially generated millimetre radiation to illuminate the scene, and here the image so formed is dominated by the reflectance distribution, which is sensitive to the orientation of reflecting surfaces and their diffusivity. Thus, active and passive millimetre-wave imaging modalities are complementary: active overcoming the low-intensity (and therefore low signal-to-noise) and 'cold sky' variability; and passive giving radiometric contrast, rather than reflectance contrast, images, without specular reflections and 'glinting' that affects the active modality.

A prerequisite of any image-based screening system is the ability to form an image in which there exists sufficient contrast between concealed objects and the background medium of the human body that they can be distinguished. This condition requires that the portion of the electromagnetic spectrum utilized for image formation is not significantly attenuated by clothing materials, and also that the objects to be resolved have physical properties: emissivity and reflectance that are sufficiently different from that of the human body to give a resolvable difference in contrast in the formed image. Also important is that the imaging system has sufficient spatial resolution that the shape of the object can be resolved sufficiently well for a decision on the nature of it, or even permitting an identification to be made—allowing, for example, the user to be able to discriminate between a handgun and a mobile phone. The only region of the electromagnetic spectrum

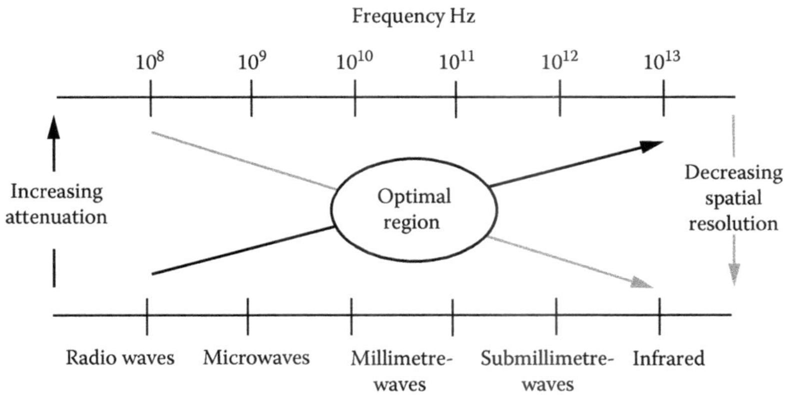

FIGURE 1.1
Illustration of the trade-off between attenuation in clothing and diffraction-limited spatial resolution over six decades of the electromagnetic spectrum; the optimal region where clothing is nearly transparent but the far-field diffraction limit is not so large as to prevent resolution of concealed objects at standoff ranges of ~10 m is found in the millimetre- to submillimetre-wave region. (From Harmer, S. W. et al., *IEEE Sensors Journal*, 99, 2013, doi: 10.1109/JSEN.2013.2273487.)

where these conditions are readily met is in that of the millimetre-wave and submillimetre-wave regime (see Figure 1.1).

In this region, imaging systems with sufficiently high spatial resolution can be achieved using the comparably small antenna sizes needed for a practical and deployable screening system. Within the shorter-wavelength portion of the millimetre-wave band and within the longer-wavelength part of the submillimetre-wave band, images may be formed through typical clothing fabrics with very little scattering or attenuation.

Most millimetre-wave imaging systems periodically sample different solid angles of the total field of view, building up the entire scene by scanning and assembling these smaller angled samples, often in the pattern of a raster scan. This can be done with a single receiving channel by scanning the directional pattern of the receiving antenna, modified by the addition of refractive or reflective quasi-optical elements, across the two dimensions of the scene, or by using multiple receiver channels and scanning spatially, thus building up multiple image pixels at every scanning position. The operation of scanning can be made redundant if a sufficient number of receiver channels are used, such that the required field of view can be met solely by the total viewing angle obtained. This approach is identical in concept to the focal plane arrays that are used to capture images in the optical and infrared bands of the electromagnetic spectrum. The disadvantages of such systems in the mm-wave and submillimetre-wave band are many. For example, there is the difficulty of positioning of a large number of receiver channels in the focal plane of the antenna with current dimensions of receiver available. Problems also arise in constructing beam-forming matrices, with spatially overlapping beams, and of implementing highly directional antennae. Furthermore, low imaging speeds and strict requirements imposed on speed parameters by the mechanical scanning of the beam steering optics limit mechanically scanned mm-wave imaging systems. The very high total cost of the systems with multiple receivers, dominated by the number of receiver channels required to achieve the required field of view, means there is a high cost associated with deploying mm-wave imaging. Another approach to provide low-cost commercial millimetre-wave imaging systems relies on using a highly dispersive antenna, in which the dispersive properties are achieved by the use of a diffraction grating and dielectric waveguide. In this design, images are formed by frequency scanning and rotation [7], [8]. These systems can form high-frame-rate images with a single millimetre-wave receiver, greatly reducing cost.

Joint use of active and passive sensors allows simultaneous formation of images of the same field of view and considerably enhances information obtainable from the scene; however, existing technical limitations prevent implementation of such systems for practical devices. Such limiting factors include complexity, high cost, comparably small fields of view, and low imaging speeds (frame rates).

Although millimetre-wave imaging systems offer distinct advantages, their large size, weight, and high cost are very significant limitations which present a barrier to more widespread uptake of the technology [9]. Currently only large businesses with stringent security requirements, for example, airports, are able to procure significant numbers of millimetre-wave imaging sensors, although this may well change with falling component costs and novel antenna designs which allow imaging millimetre-wave sensors to be realized at much reduced cost. One can envisage the installation of millimetre-wave security scanning at schools, colleges, sports events and other locations where an enhanced security presence is considered to be worthwhile. However, due to the lack of portability, millimetre-wave imaging systems will still have limited flexibility in such applications. Another approach is to utilize the millimetre- and submillimetre-wave bands with single-pixel, active, radar-like sensors, which may be used to screen people at distances from tens of centimetres to 100 metres [10]–[12]. Such sensors do not require multiple, high-sensitivity, passive receiver channels and mechanically driven optics, making them less expensive and less cumbersome than imaging sensors, but retain the ability to screen effectively under clothing. High radar resolution sensors with polarimetric capability are compact enough to be deployed as handheld, portable devices, providing a simple yes/no decision on threat level autonomously [10]. The interpretation of millimetre-wave imagery is notoriously difficult and does not lend itself well to autonomous artificial intelligence decision making; however, the interpretation of scattered radar waveforms provides a quicker and reliable discriminatory approach [13]. There are several distinct advantages that millimetre-wave radar sensors possess over their image forming analogues: spatial resolution requirements for a radar system are less stringent because imagery is not being used to identify concealed objects by their outlines; instead, the waveform of the reflected radar pulse is analysed to provide information. Consequently, standoff ranges are greater than those achievable with an imaging system because there is no requirement for multiple pixels to cover the object's angular size. Instead, illumination of the entire object is required so that all features of the object may contribute to the reflection of the incident radar waveform [10]. Operational speeds of radar sensors can be higher than for a millimetre-wave imaging system, with capture rates well above video rate possible. Radar systems are also mechanically simpler than millimetre-wave imaging cameras since there is no need for precision scanning with mechanically moved mirror; the radar sensor is a point-and-detect system. The radar system, if combined with video imagery to show the position of the millimetre-wave beam, allows simple manual aiming and screening. Size and weight savings of a millimetre-wave radar sensor over an imaging system are considerable because of the reduced number of components, optics, mechanical drives and power requirements allowing hand-held sensors to be realized. Finally, the cost of a radar sensor is significantly lower, as fewer expensive millimetre-wave components are required and expensive mechanically driven optical components are not required.

For realistic radar cross section (RCS) simulations of a human body with concealed objects, information about the permittivity of human skin and electromagtic properties of clothing and of the hidden object itself are needed. Just as important are accurate topographic specifications of the body and of the concealed object placed on the body. We have simulated the RCS of a simplified configuration, shown in Figure 1.2a, using the Integral Solver of CST Microwave Studio. The conductivity of the simulated equivalent of a human body was $\sigma_e = 0.95$ S/m and the real part of dielectric constant $\varepsilon = 12$. The material of the simulated handgun was modelled as a perfect conductor. We assume that the target is illuminated by plane wave travelling in a direction normal to the body. Figure 1.2b shows the RCS for a linearly polarized E-field aligned across the width of the simulated human body, without the handgun, while Figure 1.2c was calculated with the handgun present. We can see considereable differences in the far-field RCS patterns for both situations. However, when a real antenna illuminates the target, the situation is more complex: only a part of energy is backscattered by the concealed object, while some is reflected by the human body. Depending on the distance to the body, the beam width and position of the body, the contribution of both scattered components is highly variable, leading to a great uncertainty of the detection procedure.

The radar cross section of a person at W-band frequencies has a very large variation (>3 orders of magnitude) and can be anywhere from –30 to 4 dBsm [14] depending on the position of the person. Similarly, a handgun can present a very large RCS if the flat surfaces are perpendicular to the radar sensor,

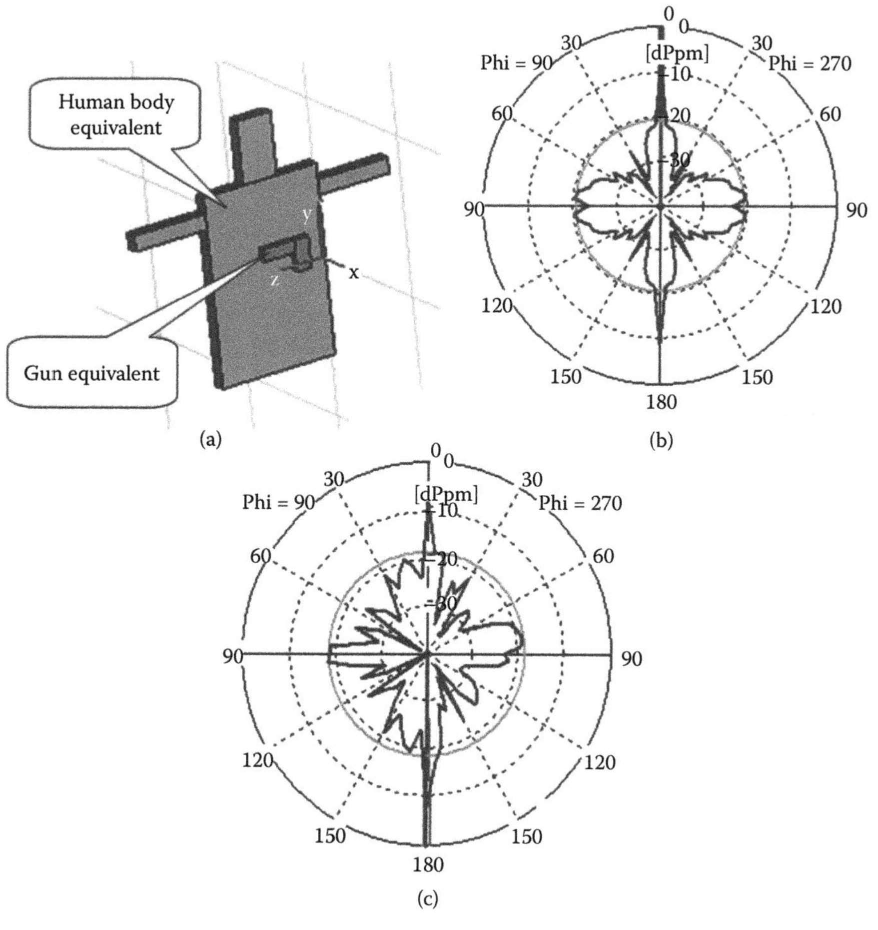

FIGURE 1.2
Simulation RCS of the human body with and without a handgun: (a) configuration of the body and handgun; (b) RCS pattern without a handgun; (c) RCS pattern with a handgun. The complex nature of the RCS with angle is evident even for the highly simplified shapes modelled. (From Kapilevich, B. and Einat, M., *IEEE Sensor Journal*, 10(11), 1746–1752, 2010. © 2010 by the Institute of Electrical and Electronics Engineers [IEEE].)

or a very low RCS if they are tilted away from the sensor. Estimating the RCS of a handgun by treating the object as a metallic cylinder (barrel) and a metallic plate (handle) gives a maximum RCS as high as ~80 dBsm, but this could drop to nearly zero for an unfavourable position. Thus, because of the very wide variability in the RCS of the human body and of typical threat objects, the detection of a weapon upon the whole body is a very difficult (if not impossible) task if one illuminates the entire human body and uses the measured RCS as the detection quantity. If the problem is constrained, for example, by having the person being screened in a pre-determined position (e.g. standing with arms by the sides and facing toward radar), then the variability in the RCS of the person can be reduced, giving a better chance to detect concealed objects by their effects on the RCS. The situation is made easier by measuring the weak polarizing effect of the body and stronger polarizing effect of typical weapons. Even so, there is still too great a variation in RCS to make such an approach a truly reliable and robust method for detection of concealed threats, with the possible exception of suicide vests, where there may be considerable amounts of metallic fragmentation, which will greatly increase the RCS in a less aspect-sensitive way than other concealed threat objects.

However, the problem of detection becomes practicably realizable when a smaller beam is used, such that the beam size is commensurate with the maximum dimension of objects to be detected. In this case the reflection of the metal surface and scattering from edges always dominates the effect of the patch of human skin which is obscured by the object, making detection by means of RCS a more robust technique. The RCS of the human torso, when illuminated by a beam which is smaller than the width of the torso, is considerably smaller and less variable than the RCS of the entire human body. Thus, the presence or absence of a threat object within this beam spot is far easier to reliably ascertain than is the case where the entire body is illuminated. The probability of detection is improved by application of polarimetric radar, and capability is further enhanced if a sufficient radar resolution can be realized such that the depth of objects present in front of the body can be also be estimated.

References

1. J. W. Lamb, Miscellaneous data on materials for millimetre and submillimetre optics, *International Journal of Infrared and Millimeter Waves*, 17(12), 1997–2034, 1996.
2. M. M. Ahlers, TSA removing 'virtual strip search' body scanners, January 2013. Available at http://edition.cnn.com/2013/01/18/travel/tsa-body-scanners (accessed 25 July 2013).
3. J. E. Bjarnason, T. L. J. Chan, A. W. M. Lee, M. A. Celis, and E. R. Brown, Millimeter-wave, terahertz, and mid-infrared transmission through common clothing, *Applied Physics Letters*, 85, 519–521, 2004.
4. M. C. Kemp, Millimetre wave and terahertz technology for the detection of concealed threats: a review, in *Proceedings of SPIE 6402, Optics and Photonics for Counterterrorism and Crime Fighting II*, 64020D, September 28, 2006. doi:10.1117/12.692612.
5. A. Aharon, D. Rozban, N. S. Kopeika, and A. Abramovich, Heterodyne detection at 300 GHz using neon indicator lamp glow discharge detector, *Applied Optics*, 52(17), 4077–4082 (2013). Available at http://dx.doi.org/10.1364/AO.52.004077.

6. L3 Security and Detection Systems. 2013. Available at http://www.sds.l-3com.com/advancedimaging/provision.htm (accessed 1 June 2013).
7. S. D. Andrenko, P. N. Melezhik, S. A. Provalov, and Y. B. Sidorenko, A planar millimeter wave antenna, in *Sixth International Kharkov Symposium on Proceedings of Physics and Engineering of Microwaves, Millimeter and Submillimeter Waves and Workshop on Terahertz Technologies (MSMW '07)*, vol. 2, 2007, p. 699. doi: 10.1109/MSMW.2007.4294782.
8. S. A. Shilo and Yu. B. Sidorenko, Millimeter wave imaging system, in *Sixth International Kharkov Symposium on Proceedings of Physics and Engineering of Microwaves, Millimeter and Submillimeter Waves and Workshop on Terahertz Technologies (MSMW '07)*, vol. 1, 2007.
9. J. J. Lynch, P. A. Macdonald, H. P. Moyer, and R. G. Nagele, Passive millimeter wave imaging sensors for commercial markets, *Applied Optics*, 49(19), E7–E12 (2010). Available at http://dx.doi.org/10.1364/AO.49.0000E7.
10. D. A. Andrews, S. W. Harmer, N. J. Bowring, N. D. Rezgui, and M. J. Southgate, Active millimeter wave sensor for standoff concealed threat detection, presented at IEEE Sensors 2013.
11. B. Kapilevich, Y. Pinhasi, R. Arusi, M. Anisimov, D. Hardon, B. Litvak, and Y. Wool, 330 GHz FMCW image sensor for homeland security applications, *Journal of Infrared, Millimeter and Terahertz Waves*, 31, 1370–1381, 2010.
12. R. J. Douglass, J. D. Gorman, and T. J. Burns, System and method for standoff detection of human carried explosives, U.S. Patent 7,800,527, 2010.
13. S. W. Harmer, N. Bowring, D. Andrews, N. D. Rezgui, M. Southgate, and S. Smith, A review of nonimaging stand-off concealed threat detection with millimeter-wave radar, *IEEE Microwave Magazine*, 13, 160–167, 2012.
14. N. Yamada, Radar cross section for pedestrian in 76 GHz band, *R&D Review of Toyota CRDL*, 39(4), 2005.
15. B. Kapilevich and M. Einat, Detecting hidden objects on human body using active millimeter wave sensor, *IEEE Sensor Journal*, 10, 11, 1746–1752, 2010.

2

Background and Theory

2.1 RCS Concept and Basic Definitions

Key physical considerations for the operation of microwave and millimetre-wave sensors form the basis for understanding the sensitivity, efficacy and limitations of such sensors in realistic environments. In the following chapters some basic concepts that relate to the physical operation of microwave- and millimetre-wave sensors are outlined.

Active sensors partially or entirely illuminate the person being screened with microwave- or millimetre-wave radiation. The amount of radiation that is reflected or scattered from the body and any items concealed thereon is determined by the radar cross section (RCS) of these objects. The RCS of easily concealed weapons, such as handguns and knives, depends strongly on the orientation of the object with respect to the transmitter and receiver as well as the frequency band of the illuminating radiation. The RCS is defined by the physical area required to intercept the transmitted intensity at the target, such that if the total intercepted power were isotropically scattered, the intensity measured at the receiver is produced. This statement clearly contains the possibility that the physical cross-sectional area the object presents to the incident radiation may not be equal to the RCS; in fact, this is generally true [1].

Mathematically the RCS can be formalized by considering the scenario depicted in Figure 2.1. An object or group of objects is located a distance R_1 from a transmitter that can transmit signals in two orthogonal states (Horizontal polarized and Vertical polarized). A receiver is located at a range R_2 from the object and is capable of receiving Horizontal polarized and Vertical polarized radiation scattered from the object(s). We assume that the beam width of the transmitted radiation is large enough that the beam entirely encompasses the object(s) and also that the intensity across the object(s) is constant. In this case the power that the object(s) intercept is

$$P_{i,pol} = \frac{P_{t,pol} G_{t,pol}}{4\pi R_1^2} A_{obj} \tag{2.1}$$

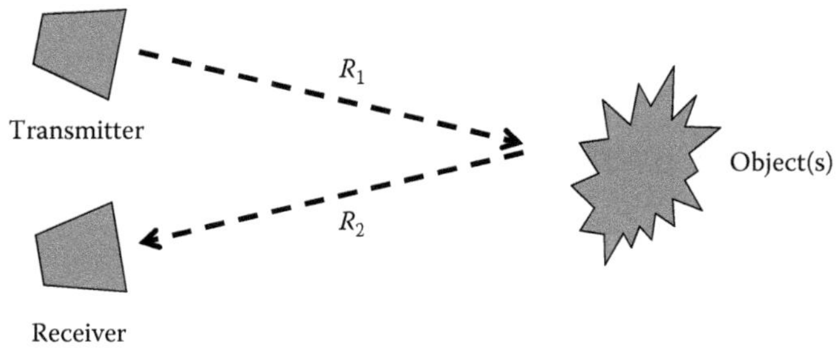

FIGURE 2.1
Scattering of radiation from a bistatic radar; the transmitter and receiver are not co-located, unlike in the monostatic case.

where the gain of the transmitter for polarization state *pol* (*pol* = *H* or *V*) is $G_{t,pol}$; the transmitted power is $P_{t,pol}$, and the cross sectional area of the object(s), in the direction of illumination, is A_{obj}. This intercepted power is scattered, in a generally anisotropic manner, and the receiver intercepts a portion of the scattered power. There is also the possibility of out-of-plane scattering. That is, the incident-polarized wave is partially scattered into the orthogonal polarization state. The effect of out-of-plane scattering is accommodated by defining the gain of the scattering object as $G_{obj,polR,polT}$. Here *polR* is the received polarization state and *polT* is the transmitted polarization state.

Using this notation, the scattered power received in the Horizontal and Vertical polarization states is

$$P_{r,H} = \frac{P_{i,H}G_{obj,H,H}}{4\pi R_2^2}A_{rec} + \frac{P_{i,V}G_{obj,H,V}}{4\pi R_2^2}A_{rec}$$

and

$$P_{r,V} = \frac{P_{i,H}G_{obj,V,H}}{4\pi R_2^2}A_{rec} + \frac{P_{i,V}G_{obj,V,V}}{4\pi R_2^2}A_{rec} \tag{2.2}$$

where A_{rec} is the area of the receiver aperture.

Using Equations 2.1 and 2.2 the results are succinctly expressed as a matrix equation:

$$\begin{pmatrix} P_{r,H} \\ P_{r,V} \end{pmatrix} = \frac{A_{obj}A_{rec}}{16\pi^2 R_1^2 R_2^2}\begin{pmatrix} G_{obj,H,H} & G_{obj,H,V} \\ G_{ibj,V,H} & G_{obj,V,V} \end{pmatrix}\begin{pmatrix} P_{t,H}G_{t,H} \\ P_{t,V}G_{t,V} \end{pmatrix} \tag{2.3}$$

From the definition of the RCS the quantity $A_{obj}G_{obj,polR,polT}$ is equivalent to the RCS of the object, which we write as $\sigma_{polR,polT}$.

Usually the gains of the transmitter for both polarization states are very similar, so that Equation 2.3 simplifies to

$$P_r = \frac{A_{rec}\sigma P_t}{16\pi^2 R_1^2 R_2^2} \tag{2.4}$$

where

$$P_r = \begin{pmatrix} P_{r,H} \\ P_{r,V} \end{pmatrix}; \sigma = \begin{pmatrix} \sigma_{H,H} & \sigma_{H,V} \\ \sigma_{V,H} & \sigma_{V,V} \end{pmatrix} \text{ and } P_t = \begin{pmatrix} P_{t,H} \\ P_{t,V} \end{pmatrix}$$

Equation 2.4 is a polarimetric generalization of the radar equation. For the sensors considered in Chapter 3, the condition that R_1 is sufficiently large that the radar beam is larger than the illuminated object (a person) is not accurate. In the case of the described radar devices, the beam is small enough that it is commensurate with the size of typical concealed weapons (~20–30 cm); this condition is optimum, as it allows estimation of the RCS of the weapon without the effect of body and background clutter dominating the measurement. Under these conditions, the majority of power incident on the person is scattered and the gain of the transmitter is only important in determining the maximum range at which the system can operate. Modification of the radar equation is straightforward since all transmitted power is incident upon the person. For a person with no concealed object,

$$P_r = \frac{A_{rec} R_{body} G_{body} P_t}{4\pi R_2^2} \tag{2.5}$$

r_{body} is the fraction of the incident power reflected from the body (the reflectance) and G_{body} is the gain of the body (a measure of the spatial distribution of the reflected power).

If the person has a highly reflective concealed object within the region illuminated by the beam, then Equation 2.5 becomes

$$P_r \approx \frac{A_{rec} P_t}{4\pi R_2^2}\left(\left(1 - \frac{A_{obj}}{A_{beam}}\right) r_{body} G_{body} + \frac{A_{obj}}{A_{beam}} G_{obj}\right) \tag{2.6}$$

Here A_{obj} is the area of the concealed object, A_{beam} is the area of the radar beam, and G_{obj} is the gain of the concealed object. A comparison of Equations 2.5 and 2.6 shows that maximum contrast between cases when an object is present and when objects are absent is obtained when $A_{beam} \approx A_{obj}$.

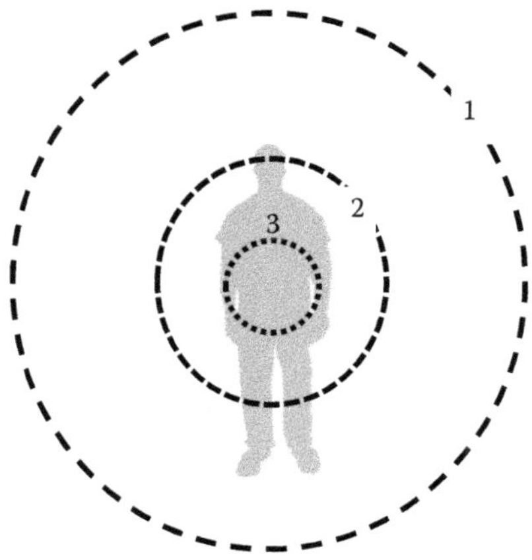

FIGURE 2.2
The radar equation (Equation 2.4) is valid for a beam indicated by circle 1; the person is significantly smaller than the beam, and the gain may be considered constant over the person. Circle 3 illustrates a beam which is entirely located upon the person, while circle 2 represents an intermediate case.

When the beam overlaps the body but is not sufficiently large that Equation 2.4 applies (see Figure 2.2), then an intermediate expression is required. In this scenario, the beam width is larger than both the body and the object, yet not so large that the gain is constant across the body. Assuming a Gaussian gain, then in the region indicated by circle 2 in Figure 2.2, the power received for a person without a concealed object is

$$P_r \approx \frac{P_t G_{body} R_{body} A_{rec}}{4\pi R_2^2}\left(1-\exp\left(-\frac{A_{body}}{A_{beam}}\right)\right) \tag{2.7}$$

where $A_{beam} = (\varphi R_1)^2$ and φ is the angular beam width. The maximum gain (on axis) is $G_0 = 4\pi/\varphi$.

In general, the RCS is strongly dependent upon the orientation of the object [2] (see Figure 2.3). The RCS plays an important role in active millimetre-wave sensors, determining the ability to discriminate concealed objects against the background body by their different scattering characteristics. The active sensors described in Chapters 3, 4 and 5 operate by transmitting an electromagnetic signal and receiving the scattered reflection from various objects present within the radar beam. As such, the RCS plays an important part in describing different objects concealed upon the human body and the ease or difficulty with which detection of these concealments can be accomplished.

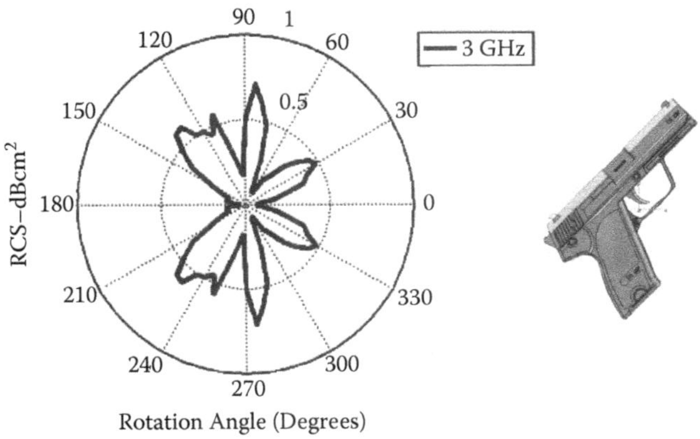

FIGURE 2.3
The calculated RCS for a handgun as a function of angle of illumination. The incident plane wave propagates from the 90° point toward the 270° point and has frequency 3 GHz. The RCS can be seen as quite a complicated function of angle with very rapid variation over small angular changes.

2.2 Active versus Passive Modes of Operation Sensors

The concept of RCS is important for understanding the physical nature of scattering of electromagnetic signals and their analysis for the purpose of object detection. These sensors are active devices and require an external source of radiation for illuminating the person who may be carrying a concealed object; passive sensors do not require an artificial source for illumination. The natural, thermal, radiation of the body and surroundings is used for the detection of concealed materials.

Active and passive sensors have inherent advantages and disadvantages relating to their effectiveness and use in different operational environments. Passive detectors rely on analysis of the thermal radiation emitted, reflected or blocked by concealed objects. Since the emitted power is very low (a receiver pointing at a blackbody at 37°C would only receive ~7 pW of radiant power per GHz of bandwidth), highly sensitive receivers are required in passive sensors. Such a receiver should be wide band in order to collect and integrate the very weak emitted signals. Therefore, the front-end circuitry of passive sensors needs to be low loss and operate over a wide frequency band. Low-noise and very high-gain wide band low-noise amplifiers (LNAs) are also required for effective amplification of the passive millimetre-wave signals.

Active detectors require a transmitter for illuminating the person being screened, and so they do not necessarily require a sensitive receiver if the source of illumination is significantly greater than the thermal background levels. Active sensors suffer from multiple reflections of the illuminating radiation, which may drastically degrade detection probability and increase false alarm rate.

The choice of operating frequency is a critical issue for both passive and active sensors. Indeed, mm-waves can penetrate to a reasonable depth through various materials (far better than infrared radiation), and have mm-wavelengths, resulting in basic resolution at the millimetre scale (better than for microwaves which have wavelengths of >1 cm). However, microwave receivers are more sensitive and demonstrate better noise performance than mm-wave receivers. Sensors operating in the THz band (submillimetre) have better resolution in comparison with their mm-wave counterparts, but they suffer from low sensitivity. As a result, THz detecting systems can operate at short distances, typically less than 1 m.

Typically, active sensors operate with coherent signals being a source of interference between several scattering layers composed of clothing, the body's skin and any concealed objects. Passive sensors are free from this effect; the noisy nature of the thermal emission signal permits considerable suppression of interference effects [3]. As a result, passive sensors can provide a higher detection probability and lower false alarm in comparable scenarios.

The effects of the environment are different for active and passive sensors. Active sensors produce returns regardless of operation in indoor or outdoor scenarios. Passive sensors are more sensitive to environmental conditions. The reason is that the passive sensor needs some thermal contrast between a background (human body, for example) and a hidden object to detect. If the difference in temperature between background and a hidden object is small enough that it is near the noise level of the sensor, then the performance of the sensor is nullified. This fact dictates positioning of a hidden object in front of a thermally emitting background. A more detailed analysis of this effect is given in Chapter 6. If a passive sensor operates in the presence of specific sources of incoherent radiation such as luminescent lamps (indoor environment) or solar emission (outdoor environment), the receiver of the passive sensor is unable to distinguish these scattering components from thermal radiation of the human body. This effect may increase false alarm rate in passive sensors.

Any practical realization of an active sensor assumes that T_x and R_x channels are inherent parts of the whole system. Isolation between both channels should be maximized to avoid degrading overall system performance. It is quite problematic to achieve good isolation if the active sensor uses the single antenna for T_x and R_x channels. The parasitic leakage between both channels is a factor in reducing the range of sensor operation (see Chapter 4 for more details) and creating an effect of parasitic return.

In order to illustrate this effect, a simulation and experiment were done using the artificial pendulum shown in Figure 2.4. It was assembled from a metal plate of the dimensions 10 × 15 cm and is free to oscillate in the direction of the arrow. The plate is positioned in front of a metal foil fixed on the

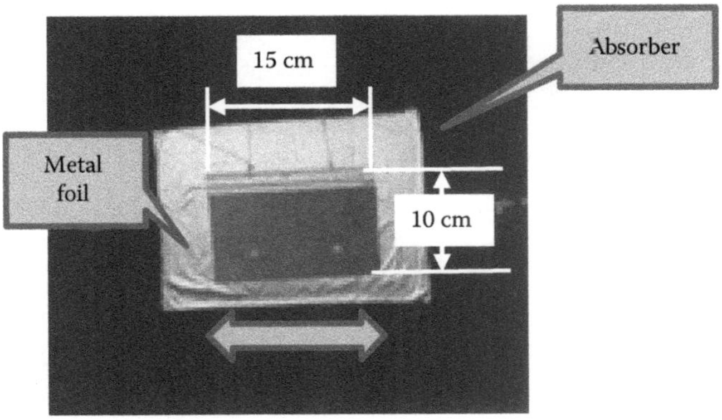

FIGURE 2.4
Artificial pendulum used in simulation and experiment.

absorbing background, resulting in the formation of two components of the scattered signal:

- *Static* return from metal foil.
- *Dynamic* return from periodically oscillating metal plate.

Now, we assume that the signal T_x is the leakage from the transmitter at the receiver's input and R_x is the return from a target with the delayed time τ:

$$T_x = A\cos(\omega_0 t) = A\cos(2\pi f_0 t)$$
$$R_x = A_r\cos(\omega_0(t-\tau)) = A_r\cos[2\pi f_0(t-\tau)] \tag{2.8}$$
$$\tau = \frac{2r(t)}{c} = \frac{2}{c}(r_0 + \upsilon_t t) = \frac{2}{c}[r_0 + D\cos(2\pi F t)]$$

Supposing for simplicity that $A_r = A$, the superposition of both signals at the input of receiver S_{in} can be written as follows:

$$S_{in} = T_x + R_x = A\left\{\cos(2\pi f_0 t) + \cos\left(2\pi f_0\left(t - \frac{2}{c}(r_0 + D\cos(2\pi F t))\right)\right)\right\} \tag{2.9}$$

where D is the amplitude of pendulum oscillation and F is the frequency of this oscillations, r_0 is static distance to the target and υ_t is its transversal velocity. Neglecting the constant phase shift term, it can be shown that superposition *static* and *dynamic* components can be presented as follows:

$$S_{in} = 2A\cos\left((2\pi f_0 t) - \frac{2\pi}{\lambda}(r_0 + D\cos(2\pi F t))\right)\cdot\cos\left(\frac{2\pi D}{\lambda}\cos(2\pi F t)\right) \tag{2.10}$$

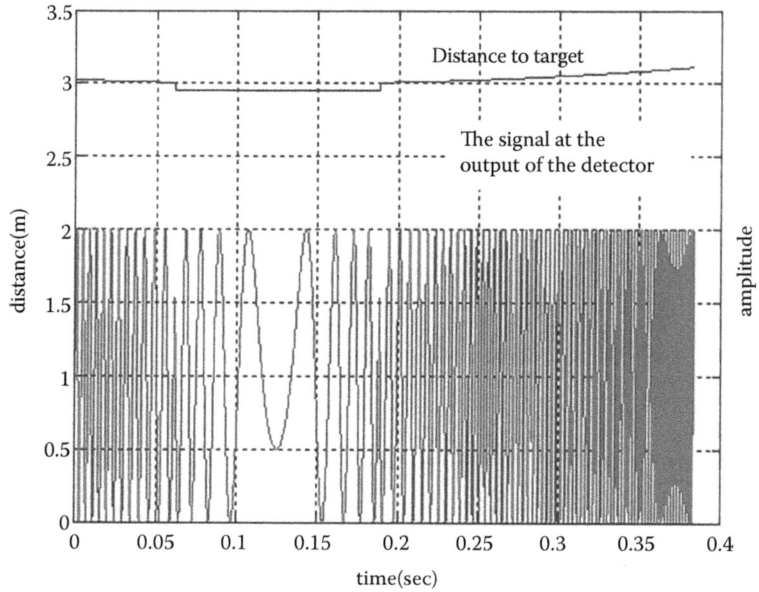

FIGURE 2.5
Example of simulation of transversally oscillating target in pendulum mode. (The amplitude is in relative units.)

Figure 2.5 shows the example of simulation of the signal after the detector's output as a function of time for $D = 2$ cm and $r_0 = 3$ m and the plot of distance to a target. There is observed a frequency chirp characterized by increasing frequency, with increased distance caused by the transversal moving of a target.

The active sensor described in [4] has been modified by replacing the two separate antennae by a single antenna and using a circulator with isolation of about 20 dB. The measured response from the oscillating target, shown in Figure 2.4 (in pendulum mode), is depicted in Figure 2.6. The experiment proves the basic effect caused by the transversal instability of a target position. The longitudinal instability of the target position also produces a chirp response. The latter is more essential for active sensors operating in the frequency-modulated continuous wave (FMCW) mode (see Chapter 4). Local instability of a target's position can be interpreted in terms of micro-Doppler effect [5], reducing an accuracy of range determination.

2.3 Millimetre-Wave Emission from a Planar Surface

There are three important properties of any real-world objects which must be taken into account when considering the operation of millimetre-wave

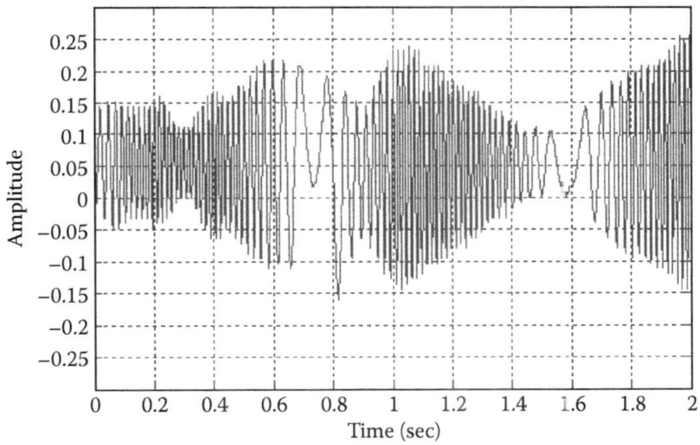

FIGURE 2.6
The measured response from the oscillating target shown in Figure 2.4. (The amplitude is in relative units.)

sensors: reflectance, transmittance and emissivity. The reflectance ρ is characterized by the ability of material to reflect a part of incident energy. This parameter varies in the range 0–1 and depends on many factors, such a physical–chemical properties, shape of object, angle of illumination, polarization, etc. In the case of a perfect reflector (non-absorber), $\rho = 1$, while for a perfect absorber or completely transparent material, $\rho = 0$. The transmittance, T, is simply the fraction of incident power that is transmitted through the object. Conservation of energy requires that $\rho + \varepsilon + T = 1$. In a real detection scenario concealed objects are typically shielded by clothing, for example, fabric or leather, meaning the reflectance is also dependent on the complex permittivity of clothing (Chapter 7). The emissivity ε is characterized by the ability of the material to emit (or equivalently to absorb) thermal radiation. This parameter varies in the range 0–1, being a function of dielectric properties, surface roughness, frequency, polarization, etc. A perfect radiator (absorber), called a blackbody, has $\varepsilon = 1$. In the millimetre-wave region of the spectrum, the emissivity of the human body is ~0.8, while that of metal is ~0.

By measuring the thermally radiated and reflected power at millimetre-wave frequencies, it is possible to ascertain information on concealed objects by a variety of methods. In the context of non-imaging sensors this can be by measuring the total power radiated into the sensor, integrated over the entire bandwidth for which the sensor operates. Using this technique, objects concealed under clothing are apparent by their screening effect on the thermal emission from the human body and also by their reflective properties. Thermal millimetre-wave emissions from surrounding materials are reflected by the human body, and this reflectance is altered by the presence of materials placed between the surroundings of the human body

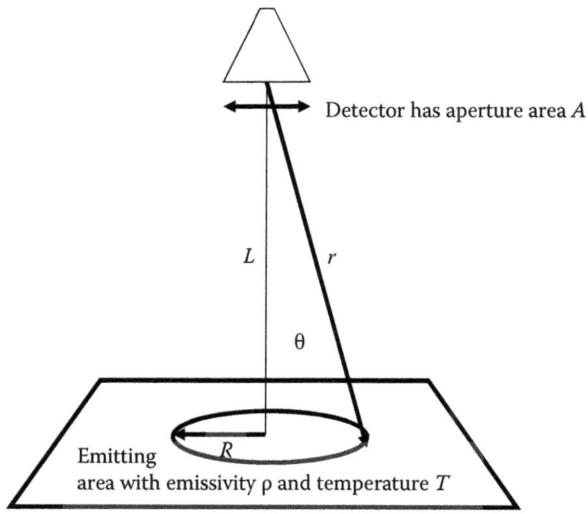

FIGURE 2.7
Sensor geometry for reception of thermally emitted millimetre-wave radiation.

and the sensor. Alteration of the expected emission from a human body without concealed objects and with expected clothing materials [6] can be measured and used to inform a threshold level; power emissions greater or less than this expectation value may then be used to predict the presence of a concealed object between the sensor and the human body. For example, a concealed metallic object will typically reduce the received power because the object will screen the typically larger thermal emissions from the body while reflecting the lower-intensity emissions from the surroundings, while a dielectric object may increase or decrease the power received depending upon the thickness and complex permittivity of the dielectric material and viewing angle.

Wien's Law shows that the peak thermal emission from the human body occurs at a wavelength of 9.5 μm [7]. The radiating power falls off rapidly toward the mm-wave region; however, the sensitivity of modern mm-wave receivers is sufficient to meaningfully measure the low-power thermal radiation and so use it for detection and imaging purposes. The efficiency of the detection depends also on the type of a target's material. Metallic objects need a temperature contrast between the human body (about 37°C) and background temperature (about 25°C) for detection, as they have a very low emissivity (~0) and a high reflectance (~1). Nonmetallic objects such as plastics, explosives and ceramics partially block the emission of the body, leading to a reduction in overall radiated power [8].

The detector only receives radiation from a small region of space, of solid angle (see Figure 2.7):

$$\Delta\Omega \approx \psi^2 \tag{2.11}$$

where ψ is the beam width of the sensor and is related to the aperture area, A, of the sensor by

$$\psi \approx \frac{c}{\nu\sqrt{A}} \tag{2.12}$$

From the sensor geometry shown, we have $r\cos\theta = L$ and $\tan\theta = \frac{R}{L}$.

The power at the receiver is given by

$$dP = \rho\beta(T,\nu)\cos(\theta)\,d\Omega\,dA_e\,d\nu \tag{2.13}$$

where $\beta(T,\nu)$ is the spectral radiance and is given by

$$\beta(T,\nu) = \frac{2h\nu^3}{c^2}\frac{1}{e^{\frac{h\nu}{kT}}-1} \tag{2.14}$$

which for $kT >> h\nu$ (which is a valid approximation in the mm-wave region for bodies at ~300 K), can be written as

$$\beta(\nu,T) \approx \frac{2kT\nu^2}{c^2} \tag{2.15}$$

For a thin circular ring or radius dR the emission area is

$$dA_e = 2\pi R\,dR$$

Hence, assuming Lambertian emission from the surface, the power at the sensor can be approximately written as

$$dP = 2\pi\rho\beta(T,\nu)R\cos\theta\,dR\,d\Omega\,d\nu \tag{2.16}$$

Therefore, the intensity on the sensor receiver is

$$d^2I = \frac{dP}{dA_d} = \frac{2\pi\rho\beta(T,\nu)R\cos\theta\,dR\,d\Omega\,d\nu}{r^2 d\Omega/\cos\theta} = \frac{2\pi\rho\beta(T,\nu)R\cos^2(\theta)\,dR\,d\nu}{r^2}$$

where

$$dR = L\sec^2(\theta)\,d\theta$$

and hence

$$dI = \pi\rho\beta(T,\nu)\sin(2\theta)d\theta d\nu$$

so that

$$dI = \pi\rho\beta(T,\nu)d\nu\int_{0}^{\frac{\psi}{2}}\sin(2\theta)d\theta$$

Giving an intensity of

$$dI = \frac{\pi\rho\beta(T,\nu)\psi^2 d\nu}{4} \tag{2.17}$$

the power received by the detector is then approximately

$$dP = \frac{\pi\rho kTA\nu^2\psi^2 d\nu}{2c^2} \tag{2.18}$$

so that

$$\Delta P = \frac{\pi\rho kT}{2}\Delta\nu \tag{2.19}$$

where ΔP is the power at the receiver and $\Delta\nu$ is the bandwidth over which the sensor is sensitive. Thus, at millimetre-wave frequencies, the thermally emitted power is proportional to the product of the emissivity of the surface and the temperature of the surface; this quantity is referred to as the brightness temperature and is the effective temperature that a blackbody radiator requires to produce the same thermal emission. The aperture area of the detector does not feature in Equation 2.19 because the solid angle which the detector views is inversely proportional to the product of the receiver aperture area, while the power at the sensor is proportional to the product of the sensor aperture area and the solid angle. This is different from the situation where one has a reflective or refractive focusing system, which can receive radiation from an angular range determined by system parameters other than aperture size. In this case, the total radiation received is proportional to aperture area. However, the sensor aperture area is important for spatial resolution when one requires viewing only a small area of the body with the sensor so as to be able to detect the presence of concealed objects. If the sensor aperture is small and therefore the viewing angle is large, the range of the sensor will be restricted, as the beam will sample receive thermal emission from the body and surroundings. As an example, the power received when a 1 GHz bandwidth sensor is pointing at a blackbody at temperature 300 K is only 6.5×10^{-12} W.

2.4 Conclusion

A variety of fundamental parameters and specific limitations need to be carefully considered in order to determine the best configuration for the millimeter-wave sensors used for remote detection of hidden objects. Modelling for these parameters often can only be accurately undertaken in limited, idealized situations. So, the final overall sensor performance is usually estimated in practice by means of serial (or iterative) experimental verifications. The considerable differences in the phenomenology for active and passive millimeter-wave sensors mean that these are complementary for detection of hidden objects. In the following chapters, passive, active and non-imaging sensors are discussed, with particular emphasis on sensors that have been designed and built by the authors.

References

1. M. Skolnik, *Introduction to radar systems*, McGraw-Hill, New York, 2001.
2. N. Yamada, Radar cross section for pedestrian in 76 GHz band, *R&D Review of Toyota CRDL*, 39(34), 2005.
3. J. Polivka, Noise can be good, too, *Microwave Journal*, 47, 66–78, 2004.
4. B. Kapilevich and M. Einat, Detecting hidden objects on human body using active millimeter wave sensor, *IEEE Sensor Journal*, 10(11), 1746–1752, 2010.
5. J. A. Nanzer, *Microwave and millimeter-wave remote sensing for security applications*, Artech House, Norwood, MA, 2012.
6. B. Lee, *Theoretical prediction and measurement of the fabric surface apparent temperature in a simulated man/fabric/environment system*, Report DSTO-TR-0849, DSTO Aeronautical and Maritime Research Laboratory, Australia, 1999.
7. Black-body radiation, http://en.wikipedia.org/wiki/Black-body_radiation.
8. R. W. McMillan, N. C. Me, D. D. Ferris Jr., and M. C. Wicks, Concealed weapon detection using microwave and millimeter wave sensors, in *Proceedings of Microwave and Millimeter Wave Technology Conference (ICMMT '98)*, pp. 1–4, 1998.

3

Active Millimetre-Wave Sensor Using Direct Detection Approach

3.1 Direct Detection Radar Principles and Theory

A direct detection receiver can be used to realize a very simple radar system. The transmitted waveform scatters from objects in the path of the radar beam and the power incident upon the receiver is measured directly, without any mixing with the transmitted waveform or a local oscillator operating at another frequency. The information retrievable from a direct detection receiver configuration is, however, limited when compared to a radar system in which both amplitude and relative phase are measured [1]. In a direct detection radar system absolute ranges to scattering objects cannot be recovered, but the axial separation between objects can be determined [2], allowing limited information about the electromagnetic structure of the scattering target to be obtained (see Figure 3.1). It is the relative simplicity of constructing ultra-wide band (UWB) direct detection radar systems allied with their lower cost that makes the use of such radar systems appealing for the application of concealed threat detection [3].

The received signal of a radar system illuminating multiple objects, as illustrated in Figure 3.1, is given by Equation 3.1, where A_R is the area of the receiver, P_T and G_T are the transmitted power and gain of the transmitter, respectively, σ_m and z_m are the radar cross section of the mth scattering target and the range to that target, respectively, ν is the frequency of the transmitted radiation, and $\Im^{-1}$ is used to denote a one-dimensional inverse Fourier transform operation. Note, it assumed that the gain of the transmitter is not strongly dependent on frequency, over the operating frequency range of the transmitter, and that the incident wave is planar across the diameter of the radar receiver.

$$S(t) = \frac{\sqrt{A_R G_T}}{4\pi} \sum_{m=1}^{M} \Im^{-1} \left\{ \frac{\sqrt{P_T(\nu)\sigma_m(\nu)}}{z_m^2} \exp\left(-i\frac{4\pi\nu}{c} z_m\right) \right\} \tag{3.1}$$

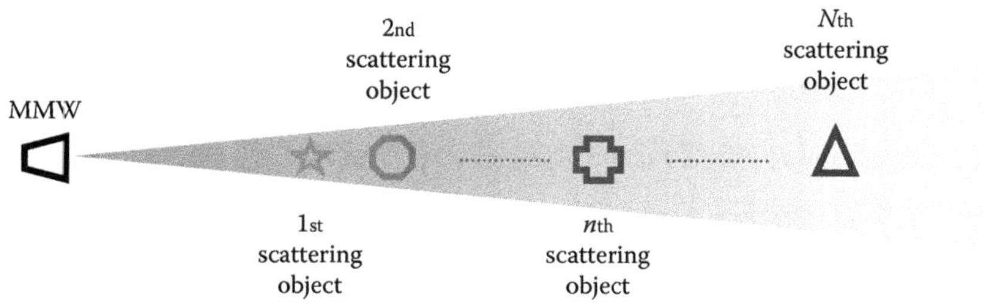

FIGURE 3.1
An illustration of a direct detection mm-wave radar illuminating several independent objects or targets at different ranges from the radar system. (From Harmer, S. W. et al., *Progress in Electromagnetics Research Letters*, 39, 37–47, 2013, doi: 10.2528/PIERL13012508. Reproduced courtesy of The Electromagnetics Academy.)

By applying the convolution theorem to Equation 3.1, information on the scattering targets illustrated in Figure 3.1 becomes clearer:

$$S(t)=\frac{\sqrt{A_R G_T}}{16\pi^3}\sum_{m=1}^{M}\frac{1}{z_m^2}\Im^{-1}\left\{\sqrt{P_T(\nu)}\right\}\otimes\Im^{-1}\left\{\sqrt{\sigma_m(\nu)}\right\}\otimes\delta\left(t-\frac{2z_m}{c}\right) \quad (3.2)$$

Therefore, we would expect to see pulses of returned signals at time intervals according to the round-trip time to that scattering target and amplitude scaled according to the radar cross section of the individual scattering object and inversely proportional to the square of the object's range. We can also see that received pulses will be broadened by convolution with the term $\Im^{-1}\left\{\sqrt{P_T(\nu)}\right\}$, which contains the transmitted power as a function of frequency, and $\Im^{-1}\left\{\sqrt{\sigma_m(\nu)}\right\}$, which is the frequency dependence of the radar cross section of the scattering objects. The transmitted power term can be compensated for [4], as this can be measured a priori; the frequency dependence of the radar cross section contains information on the axial size and structure of the scattering object. This effect is considered in greater detail later in the chapter.

Thus, from the received signal the relative positions and size of the scattering objects can be ascertained. Of course Equations 3.1 and 3.2 are only valid for a radar system in which both the phase and amplitude of the scattered electromagnetic fields are measured. If only the power (proportional to the square of amplitude) is measured, as is the case for a direct detection radar system, then information on the absolute positions and sizes of the scattering objects is lost; however, some useful information on the spacing between the objects still remains [5]–[10]. In order to obtain the equivalent signal for the case of direct detection, we simply take the absolute magnitude of the terms in Equation 3.1:

$$S(t)=\frac{A_R G_T}{64\pi^4}\sum_{m=1}^{M}\sum_{n=1}^{M}\frac{1}{z_m^2 z_n^2}\Im^{-1}\left\{P_T(\nu)\right\}\otimes\Im^{-1}\left\{\sqrt{\sigma_m(\nu)\sigma_n^*(\nu)}\right\}\otimes\delta\left(t-\frac{2|z_m-z_n|}{c}\right) \quad (3.3)$$

The signal from a direct detection radar gives a series of pulses which are separated only by the difference between any one scattering object and all other objects, while the radar cross section terms are also irrevocably mixed with each other. A direct detection radar therefore gives less information than a conventional radar system in which both the relative phase and amplitude of the received electromagnetic field are measured. With direct detection radar, the relative strengths and relative positions of scattering objects in the radar beam are lost. The separations between one object and all other objects are, however, contained within the scattered signal (see Equation 3.3).

However, although there is less information available from direct detection radar, the received signal can still be used to detect the presence or absence of objects positioned in front of the human body but concealed under clothing. In this case, there will be a double peak separated by time equivalent to the round-trip time between the front surface of the concealed object and the skin on the body. This effect is particularly useful for detecting the presence of dielectric layers of material such as wax, which are often used to simulate explosive materials [11]–[13], located between the human body and the radar sensor. Additional information, about the internal shape and structure of the target, derived from the frequency dependence of the radar cross section terms in Equation 3.3, may also be obtained if the temporal resolution of the radar is sufficient to allow such details to be discriminated from the overall scattered signal [3]. Clearly a radar sensor system for concealed weapons detection will require excellent range resolution in order to provide the capability to discern objects that are in close axial proximity.

Thin-film effects from interference within a low-loss transparent target object, for example, a sheet of wax, provide information on the product of the thickness and dielectric constant (optical depth) of the sheet. In a direct detection radar sensor, the transmitted pulse is swept, linearly with time, over its entire operating frequency range, and the received waveform is synchronized to that transmitted by, a priori, knowledge of the sweep time, allowing the appropriate frequency channels, or bins, to be allocated [11]. Transformation to the optical depth domain of the target is achieved by performing an inverse Fourier transform on the binned frequency data and scaling according to the round-trip time for electromagnetic radiation propagating in free space. In the case of a layer, or layers, of dielectric materials located between the radar sensor and the human body, the signal received from a radar, where both the amplitude and relative phase of the incident electromagnetic wave are measured, is expected to have time dependence of the form given by Equation 3.4 [3]:

$$s(t) = \Im^{-1}\{A(\nu)\} \otimes \Im^{-1}\{\Gamma(\nu)\} \otimes \delta\left(t - \frac{2z_0}{c}\right) \tag{3.4}$$

z_0 is the range to the front surface of the dielectric layer; the transmitter and receiver antenna frequency response is *A*, and the structure of the target is encoded in the reflectivity term Γ, which is the ratio of the reflected and incident electric field amplitudes. Note the parameters such as antenna gain and area that appear in Equations 3.1–3.3 have been dropped, as they do not strongly influence the shape of the received signal. The analogous signal for direct detection radar is given by Equation 3.5. where $|A(\nu)|^2$ is proportional to the transmitted power and $R(\nu)=|\Gamma(\nu)|^2$ is the reflectance, being the ratio of reflected to incident power, of the dielectric layered structure.

$$s(t)=\Im^{-1}\left\{|A(\nu)|^2\right\}\otimes\Im^{-1}\{R(\nu)\} \tag{3.5}$$

Comparison of Equations 3.3 and 3.4 clearly demonstrates that the range z_0 is lost in the case of a direct detection radar system, yet after appropriate deconvolution to remove the antenna response term, there is still valuable information in the signal encoded in the reflectance term relating to the structure of the layers. If a single sheet, or block, of wax is placed in front of the body, under clothing, so as to simulate the presence of a concealed explosive device without any metallic components, a comparison can be made between the capability of conventional radar and direct detect radar to detect such a device. Using a theoretical description based on Equations 3.4–3.6 for a wax sheet in air and Equation 3.7 for an 80 mm thick wax sheet on the human body, it is apparent that the optical depth of the wax block (~120 mm) can be determined from both conventional and direct detection radar systems (see Figure 3.2). The reflectivity of the wax sheet in air is given in Equation 3.6, while the reflectivity of the wax sheet on the human body is given by Equation 3.7:

$$\Gamma(\nu)=\frac{\Gamma_{wax}\left(1-\exp(i2\kappa h)\right)}{1-\Gamma_{wax}^2\exp(i2\kappa h)} \tag{3.6}$$

$$\Gamma(\nu)=\frac{\Gamma_{wax}+\Gamma_{body}\exp(i2\kappa h)}{1+\Gamma_{wax}\Gamma_{body}\exp(i2\kappa h)} \tag{3.7}$$

where

$$\Gamma_{wax}=\frac{1-\sqrt{\varepsilon_{wax}}}{1+\sqrt{\varepsilon_{wax}}}$$

and

$$\Gamma_{body}=\frac{\sqrt{\varepsilon_{wax}}-\sqrt{\varepsilon_{body}}}{\sqrt{\varepsilon_{body}}+\sqrt{\varepsilon_{wax}}};$$

the thickness of the wax layer is h and the value of ε_{wax} over the microwave frequency range, 14–40 GHz, over which the problem is modelled,

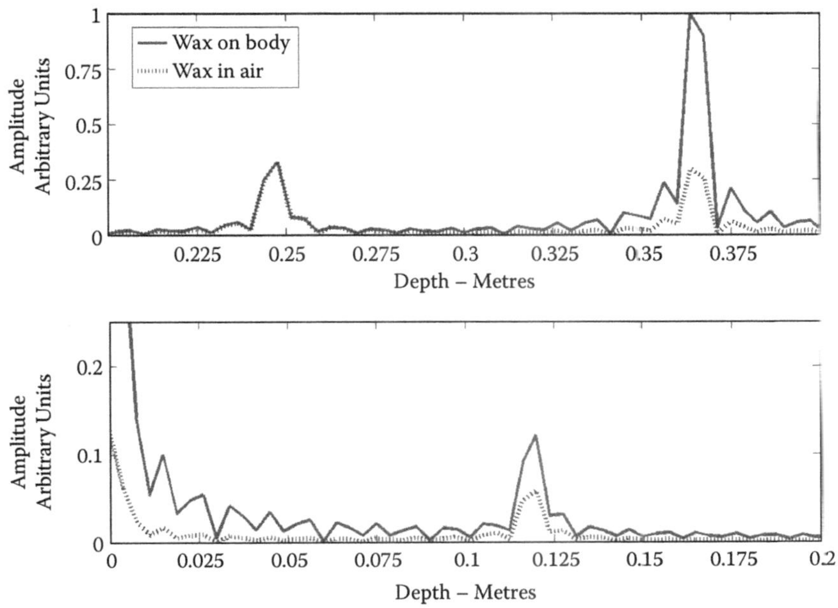

FIGURE 3.2
Theoretically predicted reflected signal from an 80 mm thick wax sheet in air and in contact with the human body. The body is modelled using measured permittivity values [14]. The wax sheet is 250 mm from the radar; 1024 frequency data points are used and the radar operates over a 15–40 GHz frequency band. The conventional radar signal (top plot) and direct detection radar (lower plot) depict the reflection from the two interfaces of the wax sheet, which are ~120 mm apart due to the optical depth of the sheet. (From Harmer, S. W. et al., *Progress in Electromagnetics Research Letters*, 39, 37–47, 2013, doi: 10.2528/PIERL13012508. Reproduced courtesy of The Electromagnetics Academy.)

is approximately constant with a real part of permittivity of 2.20 [13]. The wavenumber

$$\kappa = \frac{2\pi\nu}{c}\sqrt{\varepsilon_{wax}}$$

and the imaginary part of κ can be neglected, as it is very much smaller than the real part by virtue of the low-loss nature of wax at microwave frequencies, loss tangent of $<3 \times 10^{-4}$. The human body is modelled by the frequency-dependent complex permittivity, ε_{body} [14]. The existence of the two peaks visible in Figure 3.2. becomes clear when Equations 3.6 and 3.7 are expanded out as an infinite series of terms, the first two of which are, respectively,

$$\Gamma(\nu) = \Gamma_{wax} - \Gamma_{wax}(1 - \Gamma_{wax}^2)\exp(i2\kappa h) \tag{3.8}$$

$$\Gamma(\nu) = \Gamma_{wax} + \Gamma_{body}(1 - \Gamma_{wax}^2)\exp(i2\kappa h) \tag{3.9}$$

For a conventional radar system, the time domain signals associated with the reflectivity term in Equation 3.4 are, for the case of a wax sheet in air and for a wax sheet on the human body, respectively,

$$s(t)=\Gamma_{wax}\delta\left(t-\frac{2z_0}{c}\right)-\Gamma_{wax}\left(1-\Gamma_{wax}^2\right)\delta\left(t-\frac{2}{c}\left(z_0+\sqrt{\varepsilon_{wax}}h\right)\right) \tag{3.10}$$

$$s(t)=\Gamma_{wax}\delta\left(t-\frac{2z_0}{c}\right)+\Gamma_{body}\left(1-\Gamma_{wax}^2\right)\delta\left(t-\frac{2}{c}\left(z_0+\sqrt{\varepsilon_{wax}}h\right)\right) \tag{3.11}$$

The equations analogous to 3.10 and 3.11 for direct detection radar are given by Equations 3.12 and 3.13, respectively. The use of τ instead of t discriminates between the case where reflections are not arranged in chronological order for direct detection, unlike reflections in conventional radar.

$$s(\tau)=\left|\Gamma_{wax}\right|^2\left(1+\left|\left(1-\Gamma_{wax}^2\right)\right|^2\right)\delta(\tau)-2\left|\Gamma_{wax}\right|^2\left(1-\Gamma_{wax}^2\right)\delta\left(\tau-\frac{2\sqrt{\varepsilon_{wax}}h}{c}\right) \tag{3.12}$$

$$s(\tau)=\left(\left|\Gamma_{wax}\right|^2+\left|\Gamma_{body}\left(1-\Gamma_{wax}^2\right)\right|^2\right)\delta(\tau)+2\,\mathrm{Re}\left\{\Gamma_{wax}^*\Gamma_{body}\left(1-\Gamma_{wax}^2\right)\right\}\delta\left(\tau-\frac{2\sqrt{\varepsilon_{wax}}h}{c}\right) \tag{3.13}$$

Conversion between time and optical path length in vacuum is given by the relationship $z = ct/2$ for conventional radar signals and $z = c\tau/2$ for the depth in direct detection radar signals.

Comparing Equations 3.10 and 3.11 for a conventional radar with Equations 3.12 and 3.13 for direct detection radar reveals the characteristic optical thickness between the air-wax and wax-body interfaces, $\sqrt{\varepsilon_{wax}}h$, is present in both cases, but that range z_0 to the target is absent from the direct detection signal. Also, the amplitudes of the reflected signals at each interface are different because the strengths of reflection from the two interfaces are combined, in the sense that the relative reflectances of each interface are not temporally separable, as they are in the conventional case. It is apparent, however, that information about the optical depth of the wax sheet is still present in the signal with direct detection radar. In the example given in Figure 3.2, the bandwidth of the system is 26 GHz, giving a depth resolution (step size of horizontal axis) of no better than 6 mm in air and ~4 mm within the wax sheet with permittivity 2.20. The depth resolution is given by Equation 3.14:

$$\Delta z\geq\frac{c}{2\times\sqrt{\varepsilon}BW} \tag{3.14}$$

BW is the frequency range over which the radar operates, and ε is the permittivity of the layer.

Using an Agilent E8363B Vector Network Analyser (VNA) to take experimental results which emulate conventional and direct detection radar is valuable to demonstrate the actuality of remote sensing of concealed dielectric layers and microwave- and millimetre-wave (mm-wave) frequencies.

Two waveguide horns were attached to ports 1 and 2 of the VNA, and both horns were directed toward the target consisting of either a wax sheet in air or a wax sheet placed on the human body at a range of ~25 cm. The phase and amplitude of the scattered field were recorded with 1024 equally spaced frequencies over the band 15–40 GHz. The frequency data were transformed to the time domain using an inverse fast Fourier transform (IFFT) implemented in MATLAB®, and the amplitude of the reflected signal presented as a function of range (see Figures 3.3–3.7). Simulation of direct detection radar was achieved by multiplying the scattered frequency domain data by its complex conjugate prior to the IFFT to give a depth spectrum.

The reflected signal from a stationary human body, illuminated in the abdominal region, consists of a simple peak in the plot located at the range of the body from the radar (see Figure 3.3). The direct detection radar signal gives a peak at zero depth, as would be expected. In contrast, the signal from an isolated, 80 mm thick wax block (see Figure 3.4) shows two distinct peaks separated by 120 mm in range, while the direct detection plot has an additional peak at the expected depth of 120 mm.

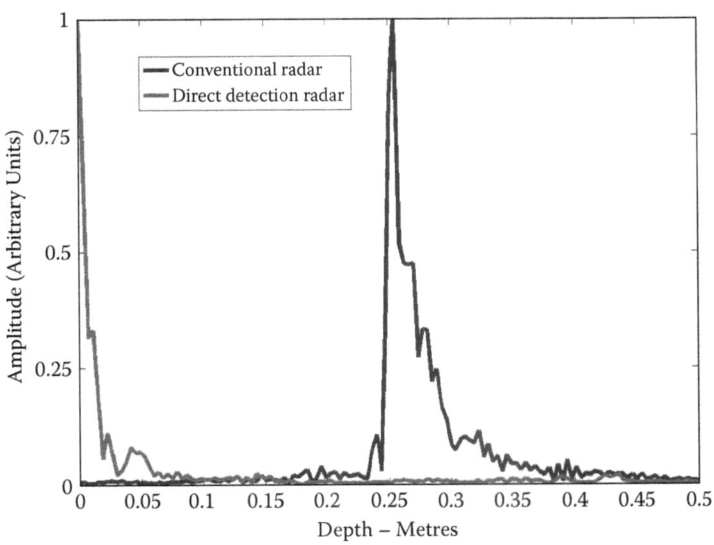

FIGURE 3.3
A graphical comparison of conventional and direct detection radar data for a clothed human torso without additional dielectric layers present; there is one obvious peak in both plots that corresponds to the reflection from the body. (From Harmer, S. W. et al., *Progress in Electromagnetics Research Letters*, 39, 37–47, 2013, doi: 10.2528/PIERL13012508. Reproduced courtesy of The Electromagnetics Academy.)

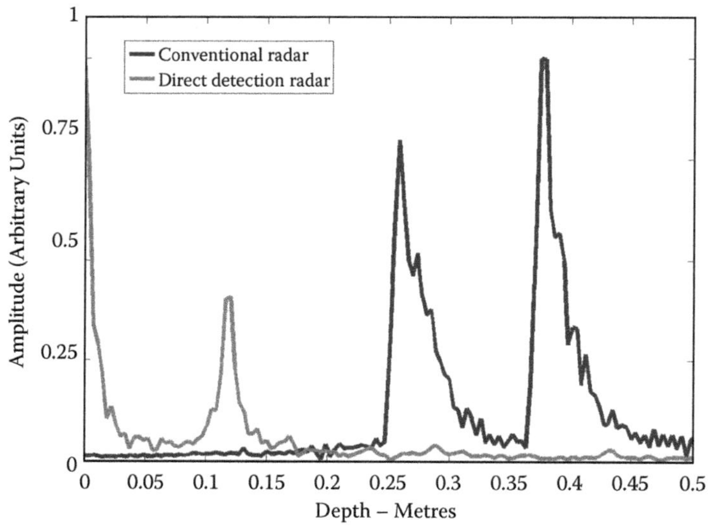

FIGURE 3.4
Signals for conventional and direct detection radar data. The target is an 80 mm thick wax sheet in air. Clearly visible are the two peaks in the conventional radar data that correspond to the reflections from the front and back interfaces of the wax sheet. In the direct detection data a peak is also clearly visible at 120 mm in the depth spectrum. As expected, the measured optical depth of the target is about 120 mm, corresponding to a refractive index of ~1.5 or a permittivity of 2.25. (From Harmer, S. W. et al., *Progress in Electromagnetics Research Letters*, 39, 37–47, 2013, doi: 10.2528/PIERL13012508. Reproduced courtesy of The Electromagnetics Academy.)

When the wax block is located immediately in front of the body's abdomen, a similar pattern is observed (see Figure 3.5), although the reflection from the body is larger than that from the wax-air interface, as there is a far greater difference between the permittivity of wax and that of the body. With a much thinner wax sheet of only 20 mm thickness the peaks are closer together (~30 mm), but are still well resolved (see Figure 3.6).

The peaks in the direct detection plot appear to be somewhat better defined than the corresponding conventional radar peaks. The data shown in Figures 3.3–3.6 have not been corrected for the electromagnetic response of the antennae, which affect the observed signal as outlined in Equations 3.4 and 3.5, and the effect is to produce a broadening or tail to the range response. Direct detection, however, is not sensitive to the phase, but only to the square of the complex amplitude; therefore, the broadening effect is reduced because the width of the temporal antennae response is shortened.

Ultra-wide band direct detection radar is able to determine the internal structure of semi-transparent materials, such as layers of explosive. However, unlike conventional radar systems, where relative phase and amplitude are recorded, direct detection radar only records the power at the sensor. Because of the simplicity of the direct detection radar (see Figure 3.7), ultra-wide band radar systems can be simply constructed using a direct detection approach,

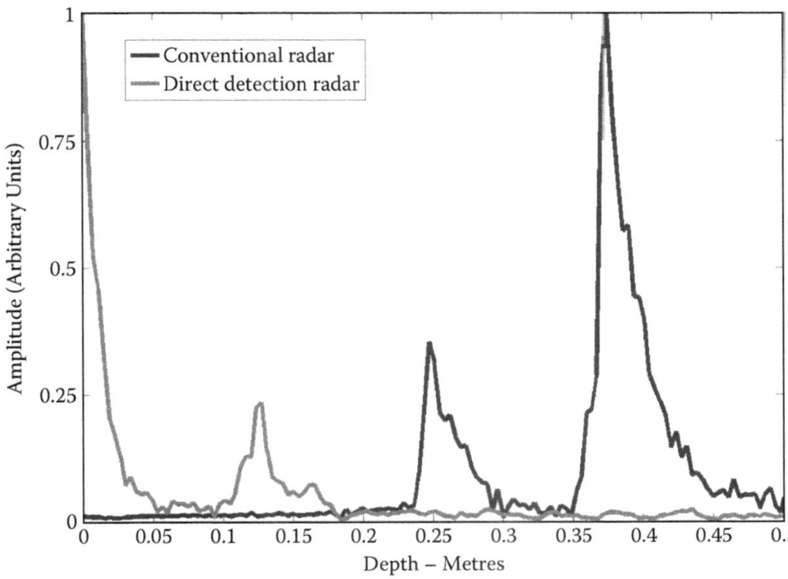

FIGURE 3.5
Conventional and direct detection radar data for a clothed human torso with an 80 mm thick wax sheet in contact with the human abdomen. Both sets of radar data indicate the presence of a target with an optical depth of ~120 mm. (From Harmer, S. W. et al., *Progress in Electromagnetics Research Letters*, 39, 37–47, 2013, doi: 10.2528/PIERL13012508. Reproduced courtesy of The Electromagnetics Academy.)

as there is no need for the expensive intermediate frequency stages that are employed in conventional heterodyne radar systems.

The significant advantage of lower cost is achieved because expensive components such as mixers are not required, and the radar receiver can be constructed from a single monolithic microwave integrated circuit (MMIC) type receiver. This is important for millimetre-wave and submillimetre-wave frequencies, where transmitters typically use multipliers and sub-harmonic detection.

The effect of antenna response (see Equation 3.5) is reduced due to the phase-independent measurement technique, giving far sharper features in the time or depth domains (see Figure 3.4, for example). Another favourable feature of the phase-insensitive, direct detection radar is the immunity of the signal to target movement. The two significant disadvantages of direct detection radar are the inability of such systems to determine the range to the person being screened and the lack of range gating for clutter rejection that is inherent due to the absence of phase information. The ranging problem can be neatly overcome by generating two alternating frequency sweeps for transmission: an ultra-wide band waveform is used for the direct detection radar for determination of object structure, and a much narrower frequency sweep is used for ranging to the person being screened by homodyne mixing.

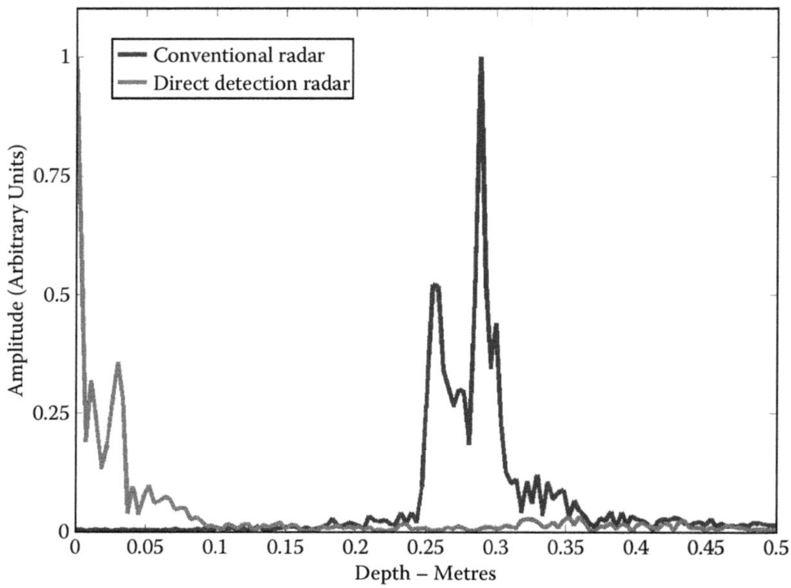

FIGURE 3.6
Conventional and direct detection radar data for a clothed human torso with a 20 mm thick wax sheet in contact with the human abdomen. Both sets of radar data indicate the presence of an object with an optical depth of 30 mm. (From Harmer, S. W. et al., *Progress in Electromagnetics Research Letters*, 39, 37–47, 2013, doi: 10.2528/PIERL13012508. Reproduced courtesy of The Electromagnetics Academy.)

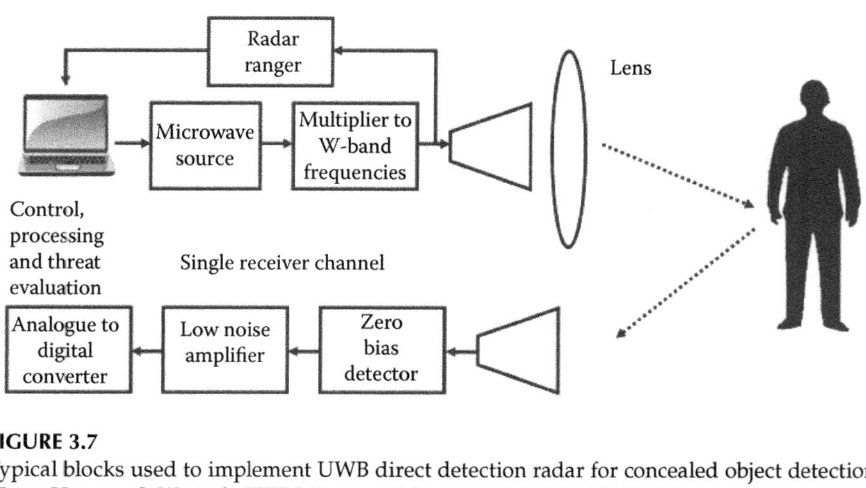

FIGURE 3.7
Typical blocks used to implement UWB direct detection radar for concealed object detection. (From Harmer, S. W. et al., *IEEE Microwave Magazine*, 13, 1, 160–167, 2012.)

3.2 Polarimetric Signatures

The two key radar principles that can be utilized for concealed object detection are the axial or depth profile information of the concealed object and polarization changes induced by the concealed object. Limited polarimetric information is obtained by transmitting millimetre radiation with the electric field aligned in one direction, for example, horizontally across the body, and measuring the scattered field both horizontally and vertically along the body using two receiver channels (see Figure 3.8). Partial reorientation of the incident, horizontally polarized, millimetre waves into the vertical polarization state is due to stronger reflection of electric field components in certain directions than in others, and is thus dependent on the object's shape and symmetry as well as its aspect with respect to the sensor. Depth profile is measured by transmission of ultra-wide band millimetre-wave radar waveforms to spatially discriminate scattering surfaces that are sufficiently separated (see Equation 3.14), as is shown in Figures 3.3–3.6. The polarizing effect of an object is often described by using a two-dimensional scattering, or Sinclair, matrix.

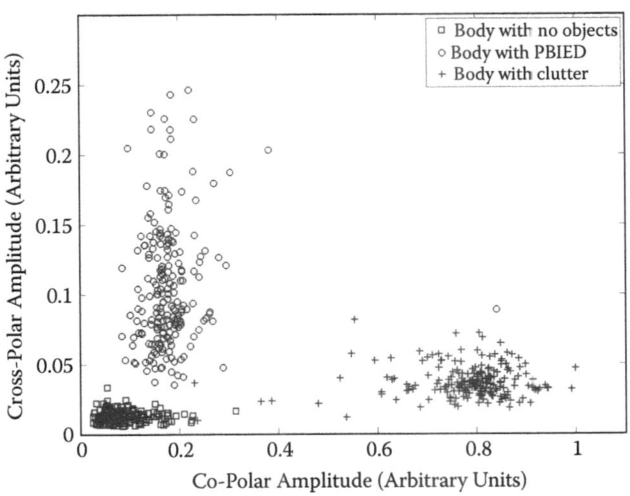

FIGURE 3.8
Co-polarized (σ_{HH}) and cross-polarized (σ_{VH}) components of a human body without any concealed objects, with typical clutter (smartphone, coins and keys) in pockets and with a mock PBIED containing fragmentation consisting of mechanical nuts and bolts. (Data taken with MiRTLE sensor at Manchester Metropolitan University by Sensing and Imaging Group. From Harmer, S. W. et al., *IEEE Microwave Magazine*, 13, 1, 160–167, 2012.)

The complex electric field amplitudes scattered from the position illuminated by a linearly polarized mm-wave radar beam are described by a 2 × 2 scattering matrix,

$$S = \begin{bmatrix} S_{HH} & S_{HV} \\ S_{VH} & S_{VV} \end{bmatrix} \tag{3.15}$$

The subscripts indicate the receiver polarization sensitivity and the transmitter polarization direction, respectively. H denotes the horizontal direction and V the vertical direction (see Figure 3.9). In conventional radar, the four components are generally complex, as each gives the fraction of the incident amplitude of the electric field reflection measured. For a direct detection radar system, the phase information is not available and the S parameters are more usefully expressed in terms of the radar cross section matrix of the object and the phase change upon reflection. Therefore, eight independent parameters describe the reflected fields from the illuminated object:

$$\begin{bmatrix} P_{R,H} \\ P_{R,V} \end{bmatrix} = \frac{G_T A_R}{\left(4\pi R^2\right)^2} \begin{bmatrix} \sigma_{HH} & \sigma_{HV} \\ \sigma_{VH} & \sigma_{VV} \end{bmatrix} \begin{bmatrix} P_{T,H} \\ P_{T,V} \end{bmatrix} \tag{3.16}$$

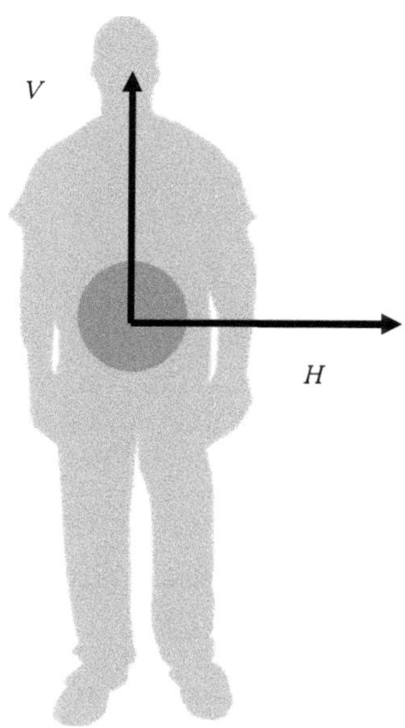

FIGURE 3.9
Illustration of the orientation of polarizations within an illuminating beam; illumination in the horizontal polarization and reception in both horizontal and vertical polarizations give sufficient information (σ_{HH} and σ_{HV}) that, when combined with depth profile data, provides a useful screening capability for detection of a wide variety of concealed threat objects [2].

The transmitted powers in the horizontal and vertical polarizations are $P_{T,H}$ and $P_{T,V}$, respectively, while the powers received in each of these polarizations are $P_{R,H}$ and $P_{R,V}$, respectively. In general, the radar cross sections are dependent upon the frequency of the transmitted radiation and may therefore give a depth spectrum as described in Section 3.1 for each of the four elements in the matrix. The radar cross section of most concealed objects is also highly aspect dependent, as is the human body, where variations of more than 20 dBsm are found at mm-wave frequencies [15].

For monostatic radar sensors, where the transmitter and receiver are co-located, or very nearly so, the off-diagonal components of the radar cross section matrix are equal. In this case we have only three independent measurements that can be made on the target at any frequency: σ_{HH}, σ_{HV} and σ_{VV}.

Small beam sizes are required to detect concealed objects reliably [11], [12]. There is a trade-off between sensor aperture size, standoff range, frequency and minimum detectable target size (see Figure 3.10). The relationship between these parameters is summarized by the expression

$$L \leq \frac{Rc}{D\nu} \tag{3.17}$$

where L is the largest linear dimension of the concealed object to be detected, R is the range to the target, c is the speed of light in vacuum, D is the diameter of the antenna aperture, and ν is the frequency of the millimetre-wave radiation. A beam with width smaller than the largest linear dimension of a concealed object will fail to illuminate the whole object uniformly. Since the radar return will not originate from the entire object, the radar cross section (RCS) of the concealed object cannot be estimated. Conversely, a beam which is much wider than the largest linear dimension of a concealed object will not localize the object sufficiently, and the radar return will contain the

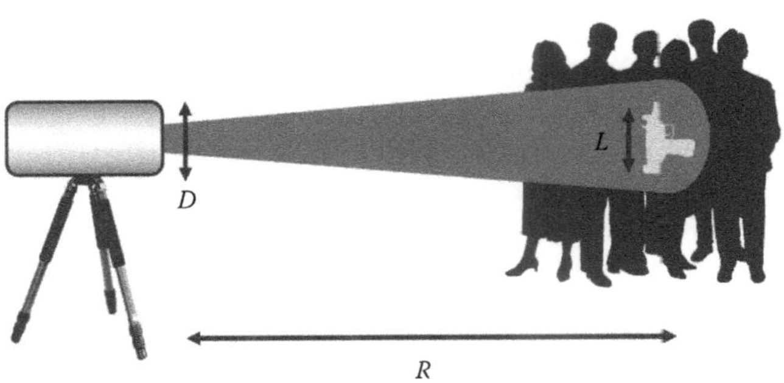

FIGURE 3.10
The millimetre-wave sensor illuminates an area on the body, which must be commensurate with the largest linear size of the object to be detected. (From Harmer, S. W. et al., *IEEE Microwave Magazine*, 13, 1, 160–167, 2012.)

signatures of other nearby items or even objects that are not located upon the body (see Figure 3.8).

In practice, L is the width of the human torso because typical handguns have a size of >20 cm, while the torso may be ~40 cm or more. Beams with wider footprints are not acceptable with direct detection radar because the incident millimetre-wave energy is partially reflected by the body being screened and partially reflected by objects around the person. In such cases the reflections from the body and the surrounding objects cannot be separated by 'range gating,' as is possible with conventional radar (see Figures 3.5 and 3.6). The effective upper range for a sensor with a 300 mm aperture operating in the W-band is ~30 m.

3.3 MiRTLE Introduction

Millimetre radar threat level evaluation (MiRTLE) is direct detection, W-band (75–110 GHz) millimetre-wave, polarimetric radar which has been developed at Manchester Metropolitan University [11], [12]. The sensor is designed to detect the presence of concealed objects upon the human body and alert the user if an object of a certain class or type is detected. For example, the system can be taught, by appropriately training a neural network, to alert the user if a handgun is concealed upon the body; yet no alert is given for typical, benign objects such as keys, coins, mobile phone handsets, etc. MiRTLE shares some common features with earlier work in the area [16]–[23], but has been primarily developed to counter the person-borne improvised explosive device (PBIED) threat at standoff ranges of up to 30 m. In operation, the system implements swept frequency continuous wave radar with low-cost components to deliver a man-portable, high-resolution (~10 mm optical depth) direct detection radar sensor capable of discriminating, and to some extent identifying, a wide spectrum of concealed threat objects. The threat evaluation procedure is made autonomously by application of a neural network to the scattered, polarimetric, depth spectrum radar data (see Figure 3.11).

The cost of the MiRTLE sensor is kept low by using direct detection receivers. This approach is significantly less expensive than implementing a heterodyne radar which operates over the same ultra-wide frequency band. The radar cross section of a concealed object is highly dependent upon the object's aspect, and so detection and identification are optimally carried out when the person under surveillance is walking or moving, and thus is presenting a variety of object positions and aspects to the MiRTLE device.

MiRTLE was developed because of the growing requirement for technologies which can provide the capability to screen people at standoff ranges, and especially in unstructured crowd scenarios. Many different types of weapons are easily concealed under clothing to prevent visual detection, and then can

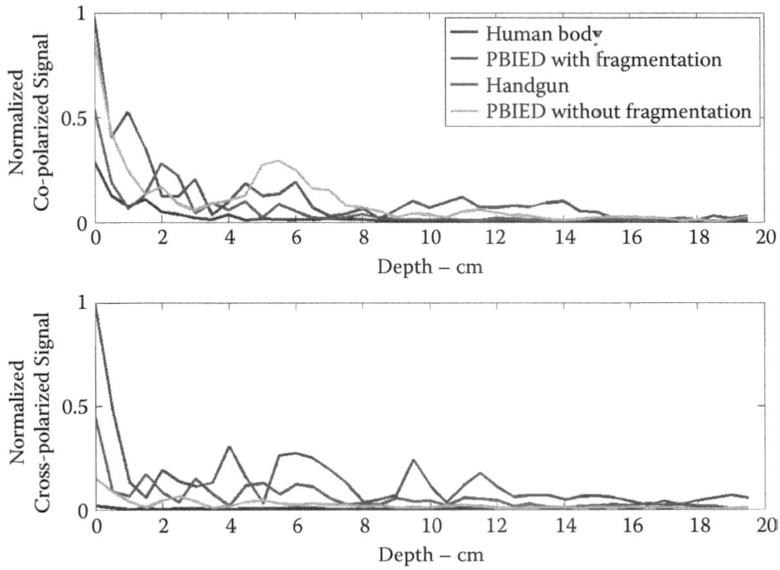

FIGURE 3.11
Measured depth spectra samples for co-polarized signal (top) and cross-polarized signal (bottom) for different concealed objects at a range of 6 m. (From Harmer, S. W. et al., *IEEE Sensors Journal*, 99, 2013, doi: 10.1109/JSEN.2013.2273487.)

be used to attack the public and target national infrastructure, giving terrorist forces a great and asymmetric advantage over security forces. Perhaps the most effective and feared terrorist weapon is the PBIED, these multifarious weapons are easily concealed under standard clothing, and detection and mitigating action are certainly required before the bomber nears the target area.

A variety of electromagnetic technologies have been developed to screen for various concealed threats, such as handguns, knives and PBIED. These sensors include portal metal detectors, trace explosive vapour detection and x-ray backscatter imagers. However, these devices are limited to fairly short-range operation and are often deployed in a portal. Millimetre-wave (mm-wave) and submillimetre-wave imaging (terahertz) is the most widely applied image-based screening technology for concealed threat detection (CTD). This is because typical clothing is highly transparent to electromagnetic radiation at mm-wave frequencies [24]–[27] and sufficient spatial resolution is available, with a practicable aperture size for identification of concealed objects at short ranges of a few metres. Objects concealed under clothing can only be resolved if they possess a different temperature, emissivity or reflectance than that of human skin, and these physical differences will manifest as contrast in the captured imagery. However, limited spatial resolution constrains the standoff ranges for effective screening with mm-wave imaging in CTD applications [28]. Mm-wave cameras also suffer from low frame rates due to the very low intensity of passive thermal mm-wave emission, and from limited fields of view because the very high cost of mm-wave receivers makes an array of mm-wave receivers prohibitively expensive. Commercial mm-wave cameras rely on mechanical scanning of reflective optics, although there are some different approaches which circumvent some of these problems [29], [30].

An alternative to mm-wave imaging for the application of CTD is UWB millimetre-wave radar. The CTD radar functions by using a directional antenna to illuminate the person being screened with UWB mm-wave radiation; the scattered waveforms are processed and analyzed to give a threat assessment. There are systems, in production or under development, that use non-imaging techniques to detect concealed threats carried on the human body [16]–[23].

The system developed at Manchester Metropolitan University [11], [12] (MiRTLE) analyzes the scattered depth domain waveforms [31]–[33] in two orthogonally polarized channels, to determine the likely nature of the target (see Figure 3.11). As an example, the depth domain waveform for scattered polarimetric radar from the human body with no concealed object present is different from that obtained with a concealed handgun and different again from that obtained with a concealed PBIED. The ability to discriminate a threat object from a non-threat object is determined by the different spatial distributions of material within certain classes of object. The polarizing effect of the object in conjunction with the object's depth spectrum is often sufficient to enable reliable detection and discrimination of concealed objects, even though these two effects depend on the orientation of the concealed object.

As an example, very few non-threat objects that are legitimately carried concealed on one's person are shaped like a handgun or consist of large amounts of metal fragments distributed over the torso and abdomen. These targets can therefore be reliably discriminated from the typical clutter of benign items carried by most people.

Figure 3.11 shows the depth spectra obtained for a gun, a block of wax (PBIED without fragmentation) and a block of wax containing fragmentation that were strapped to the front of a human body. It is apparent that the metal objects give a much increased cross-polar response when compared to the body alone. It may also be noted that when the wax blocks are present, an increased signal is also observed in the co-polar spectrum, particularly around a depth of 5 cm, which corresponds to the optical or electrical length of the block.

Mm-wave radar for CTD enjoys some significant advantages over imaging systems:

- Faster autonomous detection and discrimination of certain classes of targets, for example, PBIED fragmentation components, than is achievable with mm-wave imagery
- Less stringent requirement on spatial resolution, as beam size is required to be commensurate with that of concealed objects, ~200 mm,

and thus the range at which these system can work is much greater than is the case for imaging systems with like sized apertures

- Faster operation, as the transmitter and receivers can be easily swept at high rates
- No moving parts, unlike imaging systems which rely on flapping mirrors
- Considerable size, weight and cost savings over a millimetre-wave imaging system due to reduced number of components, optics, mechanical drives and power requirements

3.3.1 MiRTLE Sensor Design

MiRTLE transmits beams of linearly polarized W-band (75–110 GHz) millimetre-wave radiation. The focused beam has a far-field divergence of less than 1° and a directivity of ~45 dBi. The beam covers an area which is approximately the width of a human torso at a range of 25 m. The system allows the operator to perform a real-time threat level assessment of the target at approximately video frame rates by reception of the depth spectra of orthogonal polarization states, enabling distinction between different types of objects [34]. A single person may operate a unit, and the sensor is deployable at variable target ranges from a few metres up to the maximum operating range of 25 m. MiRTLE has a compact design with the transmitter and receiver modules co-located within the unit, giving essentially what is a monostatic radar system (see Figure 3.12).

Previous work in the field [9], [31], [34] has shown scanning over a very wide frequency band has the ability to give long-range depth information about the target. This is essential for detecting layers of dielectric without metallic inclusions attached to the human body and also improves identification of bulky metal objects such as handguns. To achieve sufficient radar resolution of less than 10 mm requires bandwidths of ~30 GHz. The MiRTLE system operates over the entire W-band; this provides the required bandwidth with suitably low attenuation in clothing and the ability to operate at reasonable ranges.

The direct detection receivers incorporated in MiRTLE measure the power of the returned signals only. Heterodyne receivers have the ability to measure relative phase information, and hence determine the range to the target and surrounding objects, enabling the rejection of background clutter by range gating. The capability to reject scattered waveforms from differing target ranges is an important consideration, as it is an effective method of extending the operational range of a system without increasing the ratio of aperture size to transmitted wavelength. This is because beams which are substantially larger than a person can be used to screen; reflections which do not originate from the person and their carried objects can be ignored. A direct detection receiver cannot differentiate between reflections from

FIGURE 3.12
Photograph of a prototype MiRTLE system showing the focusing lens and receiver horn antenna.

the person and those from background clutter. A heterodyne architecture, however, requires tracking the transmission and local oscillator frequencies precisely and at high speed. This added system complexity must be weighed against potential clutter rejection and sensitivity improvements as well as system cost. As was shown in Equation 3.3, the long-range structure of the target is retrievable, in a limited fashion, with direct detection by performing an inverse Fourier transform of the UWB frequency spectrum data [31].

It should be noted that when using a power-only detection system, the signals are not sensitive to Doppler smearing, which would be seen in phase-sensitive detection systems with moving targets. This is important for people, who can be moving at several ms^{-1} and would require that the radar measurement be completed before the target has moved a significant fraction of a wavelength, i.e. less than 0.1 ms at 90 GHz. Additional measurements at a series of frequencies with a phase-sensitive system to exhibit the target structure would therefore require a high sampling rate (>100 MHz). With power detection the lack of Doppler broadening permits the demands on the sampling rate to be relaxed; however, it is still advantageous to perform the radar measurement over a time during which the target signal has not significantly moved. This is much less demanding and the duration of the full sweep is about 1 ms.

A millimetre beam emitted from a 20 dB gain horn antenna is focused by a polyethylene lens to achieve an overall antenna directivity of 45 dBi. The angular beam size for an aperture of diameter D and a diffraction-limited beam is approximately λ/D. For a 300 mm diameter lens at 90 GHz, this gives an angular beam width of ~0.01 radians, which at a range of 25 m corresponds to a spot size of ~250 mm. This is comparable with the width of the human torso and also the probable size of PBIEDs. Measurements of the beam profile shows the beam to be approximately Gaussian for ranges greater than 4 m with a width in close agreement with that predicted by application of well-known far-field diffraction theory for a circular aperture. The transmitter horn antenna is located at the focus of the lens to give optimal resolution at long range. In this configuration the Gaussian beam waist is effectively at the position of the lens and the beam diameter increases continuously with distance from the lens. The beam pattern from the transmitter horn is arranged to extend slightly beyond the lens aperture to give an optimal transmitted beam profile.

To provide an ultra-wide band source which can be swept in ~1 ms, a free-running voltage-controlled oscillator (VCO) is used to sweep a microwave signal from 12.5 to 18.3 GHz; this signal is input to a six times active frequency multiplier module to give in excess of 1 mW output power over the millimetre-wave frequency range 75 to 110 GHz. A 10 dB waveguide coupler mixes the return signals, received in a confocal arrangement back through the lens and horn, with approximately −40 dB of the transmitted power onto a detector diode. This homodyne signal is amplified over a 1 MHz bandwidth and is used to acquire the range to target, which is required for scaling of the UWB polarimetric depth spectra.

The polarimetric depth spectra are obtained by two receiver modules which measure the return power through 25 dB gain waveguide horn antennae, arranged so as to be sensitive to parallel or orthogonal polarization states (with respect to the transmitted polarization). Earlier versions of the system [12] used commercial high-sensitivity MMIC detector modules developed for use in passive imaging systems, but these could not be swept rapidly enough for effective screening rates. The received millimetre-wave signals are amplified by 20 dB full-band, low-noise amplifiers (LNAs), and the microwave power detected with zero-bias diodes. This gives the benefit of a significantly flatter and broader band microwave response and also a faster detector response. The co-polarized and cross-polarized detector signals are amplified by high-input impedance operational amplifiers with a signal bandwidth from DC to 30 kHz.

A schematic outline of system components is shown in Figure 3.13. A PIC32 microcontroller board together with external, high-speed digital-to-analogue converter (DAC) and analogue-to-digital converter (ADC) is used to provide the control voltage to the VCO and to digitize the voltage outputs from the detector modules. The microcontroller board also performs an inverse fast Fourier transform (IFFT) on each signal with 2048 points for

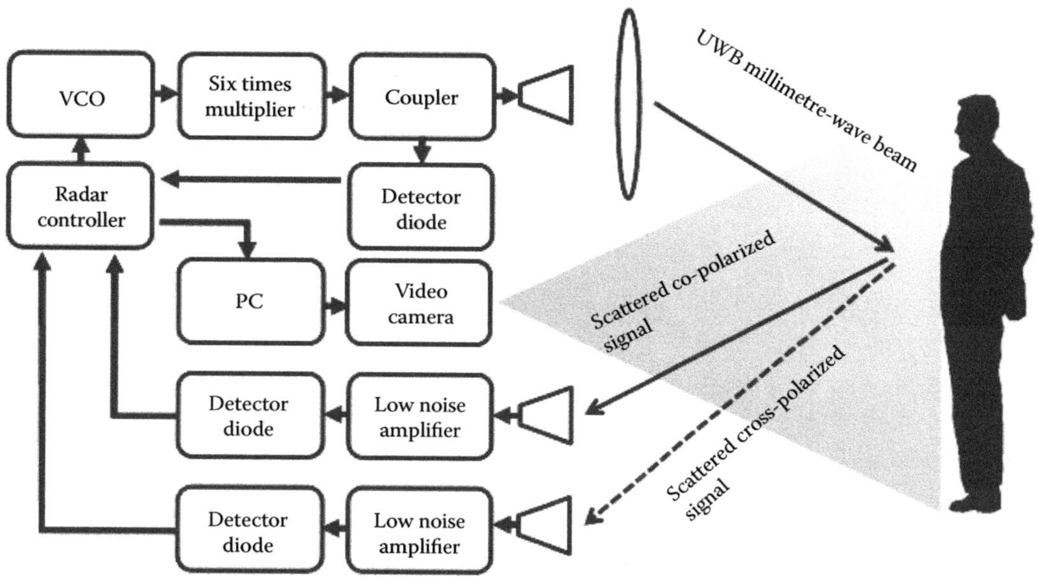

FIGURE 3.13
A block diagram showing the major sub-systems required to implement UWB, polarimetric radar in the mm-wave band with direct detection. (From Harmer, S. W. et al., *IEEE Sensors Journal*, 99, 2013, doi: 10.1109/JSEN.2013.2273487.)

range finding and 256 points for the polarimetric signals. Fewer points are required for threat detection, as the maximum likely threat size is small compared to the maximum operating range of the radar.

The range is calculated using the position of the intermediate frequency peak from the range finder signal; these data, together with the frequency spectrum of the receiver signals, are exported to a connected laptop. Each voltage sweep takes about 1.5 ms, and a packet of data consisting of a single range measurement and the mean of four successive measurements of the polarimetric signals is sent to the laptop via a USB link at up to 30 times/s.

A continually updating threat assessment is displayed on a laptop computer, which is integrated with a video feed, operating at 30 frames per second, from a USB-connected video camera. A circular cursor indicating the millimetre-wave beam position on the video image is presented to the user, who can manually scan the beam over the target. The size of the cursor is chosen to match the approximate size of the millimetre-wave beam on the target. Since the video camera is mounted on the unit, the video display, in combination with the measured target range, gives an accurate indication to the operator the position of the microwave probe beam with respect to the person. This allows the user to maintain a good beam position on the person being screened and facilitates in screening different areas on the body.

An indication of the result of the threat assessment is presented to the opertor in real time (see Figure 3.14). The indication is by audible alarm and/or a colour-coded cursor display: red for threat suspected, yellow for threat possible and green for no threat. Also displayed is the range to the target, obtained by analysis of the signal from the homodyne range finder, which gives the distance of the object illuminated by the beam up to the maximum operational range of 25 m. Beyond this distance, an out-of-range indication is displayed.

The process of scanning the person, typically standing or walking toward the system, is performed by the operator, who manually steers the direction in which the instrument is pointing. The operator uses the cursor position

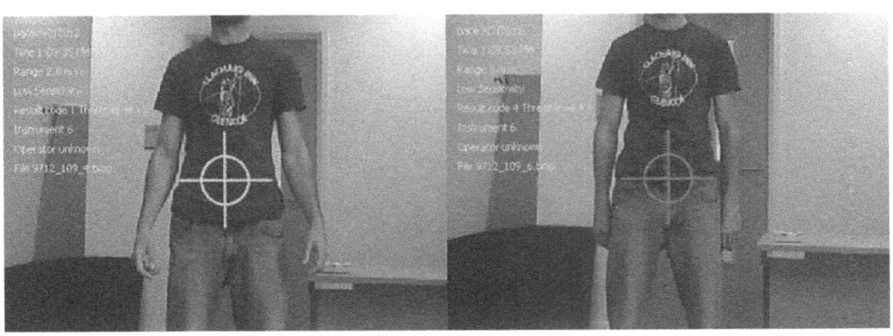

FIGURE 3.14
Cursors, used to indicate threat level assessment, superposed onto the video imagery. The cursor size is commensurate with that of the millimetre-wave beam. (From Harmer, S. W. et al., *IEEE Sensors Journal*, 99, 2013, doi: 10.1109/JSEN.2013.2273487.)

and video feed to determine which area is being scanned. The sequence of the scan will depend on operational requirements, but will typically involve quick lateral or vertical scans across the torso, followed by scans that are more detailed over areas which are indicated as a possible threat location.

3.4 Radar Signal Classification Techniques and Results

MiRTLE receives the scattered waveforms in both co-polarized and cross-polarized states with respect to the transmitted beam. The very simplest, yet reasonably effective, approach is to analyze the magnitude of the cross-polar signal only. When appropriately scaled with range and transmitted power, the cross-polarized signal gives a good indication of the amount of out-of-plane scattering which results from the object [34]. This has been found to be significantly larger for metal objects with sharp edges than for the human torso with most clothing types. Figure 3.15 shows the signal values averaged across the entire frequency sweep for a person at a range of ~10 m with and without a simulated PBIED consisting of steel nails embedded in a wax block. The device size was approximately 200 × 200 mm and 20 mm thick. The data points corresponding to with and without the device are clearly separated.

The analogous measurement made at a greater range of 20 m is shown in Figure 3.16. There is still clear distinction between the two sets of points, but now there is some overlap between the edges of the two distributions. This effect is due to several factors:

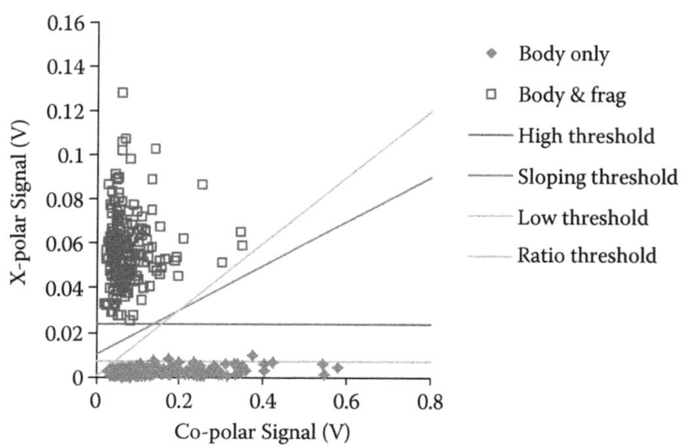

FIGURE 3.15 (See color insert.)
Measured co-polarized and cross-polarized signal amplitudes from 200 radar returns resulting from the human torso at a range of ~10 m. The person is either without objects or has a simulated PBIED. (From Harmer, S. W. et al., *IEEE Sensors Journal*, 99, 2013, doi: 10.1109/JSEN.2013.2273487.)

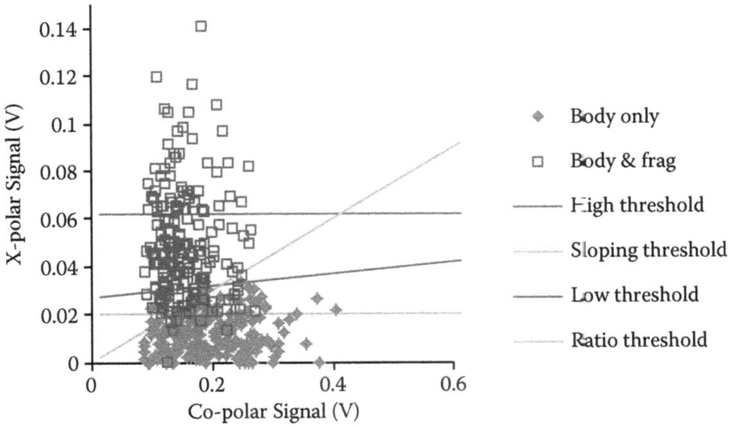

FIGURE 3.16 (See color insert.)
Measured co-polarized and cross-polarized signal amplitudes from 200 radar returns resulting from the human torso at a range of ~20 m. The person is either without objects or has a simulated PBIED. (From Harmer, S. W. et al., *IEEE Sensors Journal*, 99, 2013, doi: 10.1109/JSEN.2013.2273487.)

1. The beam size becoming considerably larger than the PBIED, and therefore diluting its response with that of the surrounding body area.
2. The beam size becoming comparable in size with the body width, and hence acquiring return signals from the edges of the body; this effect results in a higher cross-polarized signal.
3. The signals becoming weaker with increasing range of object, and hence more affected by system noise and background signals.

To investigate the influence of range upon results, the mean of the polarized signals over the operational frequency band was measured at different distances; the results are shown in Figures 3.17 and 3.18. These results are presented on a log-log plot with respect to range, R, and, for comparison, the dependence $1/R^2$ is also shown. It is apparent that the $1/R^2$ dependence describes the behaviour closely over the variation of ranges shown, although at the longest range the signal from fragmentation is beginning to fall off faster than $1/R^2$. This result is anticipated since the target is comparable with the beam size at 20 m. At much longer ranges the signals are expected to fall as $1/R^4$ as is predicted by the radar equation for objects much smaller than the beam size. However, this regime is inherently beyond the operating range of the direct detection sensor since the requirement is for the beam size to be commensurate with object size. These results suggest that a modified expression may be devised which describes the behaviour across the intermediate region, between inverse square and inverse quartic, for example, one of the form $R^{-2}(1 + (R/R_0)^2)^{-2}$. This modified dependence is shown as the dotted curve in Figure 3.17. The range parameter R_0 corresponds to a range

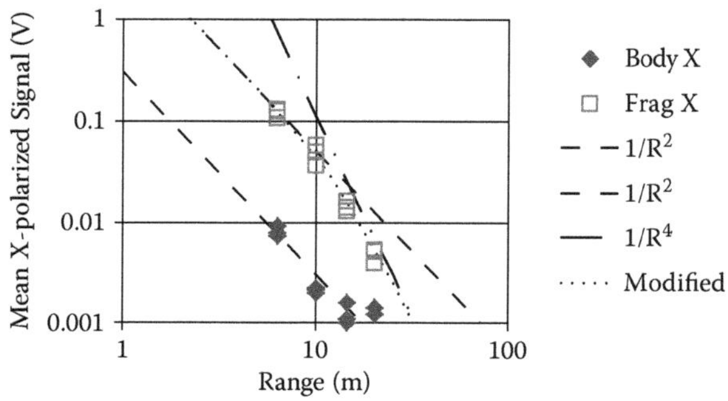

FIGURE 3.17 (See color insert.)
A log–log plot of the mean value of the measured cross-polarized signal amplitude (V) from the human body as a function of the range to the body. The body has no objects or has an attached PBIED. The inverse square range dependence is shown for comparison along with the modified expression. (From Harmer, S. W. et al., *IEEE Sensors Journal*, 99, 2013, doi: 10.1109/JSEN.2013.2273487.)

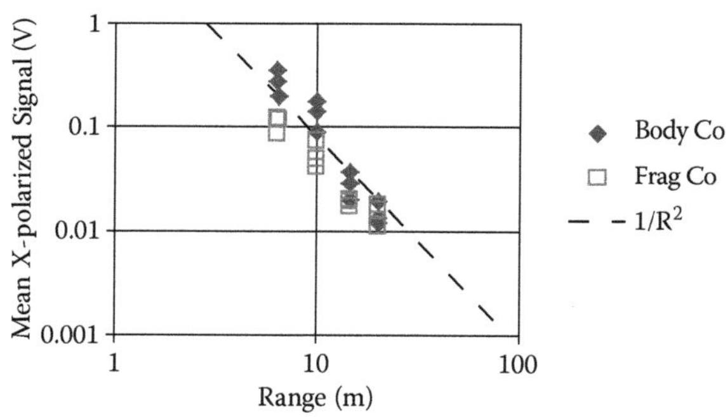

FIGURE 3.18 (See color insert.)
A log–log plot of the mean value of the measured co-polarized signal amplitude (V) from the human body as a function of the range to the body. The body has no objects or has an attached PBIED. The inverse square range dependence is shown for comparison along with the modified expression. (From Harmer, S. W. et al., *IEEE Sensors Journal*, 99, 2013, doi: 10.1109/JSEN.2013.2273487.)

~15 m where the beam size is similar in size to a typical concealed threat object, ~200 mm. The maximum operational range of the MiRTLE system is about 30 m; this limit is set by the random noise level of the system, which is less than 1 mV; the beam divergence, which is about ~1°; and the typical width of the human body, ~ 300–400 mm.

MiRTLE can be configured to alarm when the cross-polar signal is larger than a pre-determined, range-scaled threshold value equal to K/R^2. K is proportional to the expected cross-polarized radar cross section of a typical person with clutter items on his or her person. The value of K is set so that any person exhibiting a cross-polarized radar cross section which is larger than an expected value gives an alarm. This threshold is normally set quite high, so that the probability of false alarms due to noise, background and small, benign, but frequently carried metal objects is sufficiently low, yet still results in a high probability of detecting a fragmentation device. Inspection of Figures 3.17 and 3.18 clearly demonstrates that a high threshold value reduces the detection rate, especially at longer ranges.

For other, more constrained applications, e.g. when standoff screening people who should not be carrying any objects, detection of all objects may be an advantage; in this case a lower threshold may be useful. This nominal lower threshold is also shown.

Using a fixed threshold, based on the cross-polar signal, is a crude, but surprisingly effective feature for target classification. Other threshold methods can also be employed to improve object classification capability. For example, a sloping threshold, which also includes the co-polarized signal, gives an improvement in discrimination, as there is significant correlation of co-polarized and cross-polarized signal amplitude. Another method, whereby the threshold is based on the ratio of cross-polarized and co-polarized signals, also gives reasonable performance, as shown here, and has the desirable feature of not requiring knowledge of the object range, since the range dependence of the terms cancels.

Although a simple threshold discrimination approach is successful in identifying some strongly polarizing objects such as fragmentation devices, it is less successful at distinguishing smaller, smoother-sided benign objects such as cameras and mobile phones from weapons such as handguns. The results for a small camera and small handgun are shown in Figure 3.19. At a short range of 6 m it is difficult to discriminate between these different targets.

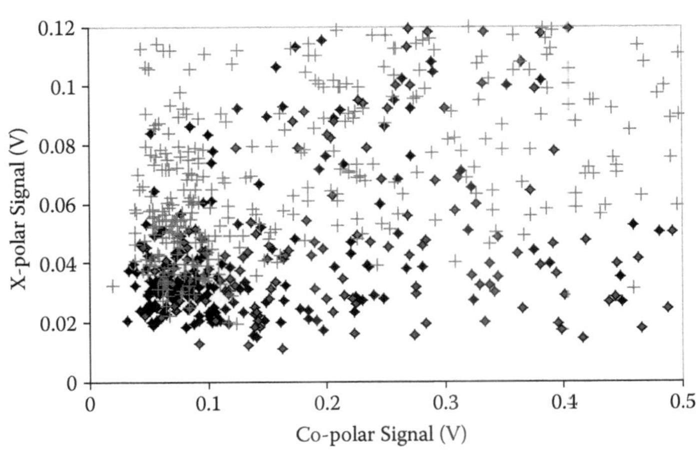

FIGURE 3.19 (See color insert.)
Measured co-polarized and cross-polarized signal amplitudes from 200 radar returns at a range of 6 m for a person with either a compact camera (diamonds) or small handgun (crosses) attached to the body. The groups of points are not clearly separated. (From Harmer, S. W. et al., *IEEE Sensors Journal*, 99, 2013, doi: 10.1109/JSEN.2013.2273487.)

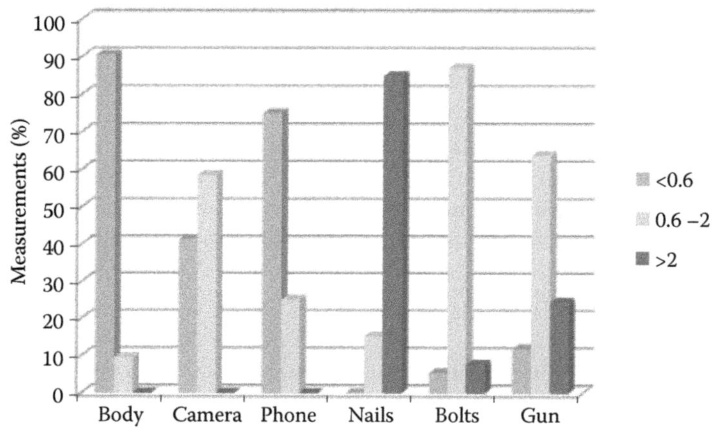

FIGURE 3.20 (See color insert.)
The percentage fractions of cross-polarized measurements which were below the lower threshold 0.6 (green), between 0.6 and 2 (yellow), and above 2 are shown for the different objects on the body. (From Harmer, S. W. et al., *IEEE Sensors Journal*, 99, 2013, doi: 10.1109/JSEN.2013.2273487.)

In Figure 3.20 the results of processing the signals with a two-level threshold discriminator are shown. With this configuration, the larger PBIED is consistently identified as being above the upper threshold, but the other weapons have a much lower probability of detection. A classifier with considerably enhanced performance, achieved by utilizing more information in the received signals, is described in the next section.

Artificial neural networks (ANNs) are software classifiers, which are well suited to situations in which the classification criteria are not known a priori, but are learned by training with signals from known objects. They have been previously used [2] for classifying radar signals in the time and frequency domains for the application of CTD. The ANN has input nodes, matched to the input data format, output nodes in the form of a threat probability, and 10 hidden layer nodes. The connections between the nodes are initially random, but are optimized by training using a set of data containing examples with and without various threat and benign objects. The network type is a feed forward back propagation type and uses sigmoid neurons. The data input is a compressed version of the inverse Fourier transform of the swept frequency spectral responses from the co-polarized and cross-polarized detectors and constitutes what the authors refer to as a depth spectrum [31]; this contains data about the size of the object in the range direction. The data are compressed by discarding data points in the IFFT of the frequency spectrum that correspond to separations greater than 20 cm; only objects with depth profiles below this size are not practical to effectively conceal under clothing.

The results for a particular case with various items strapped to the human torso are shown in Figure 3.21. Two hundred scans of each object were taken

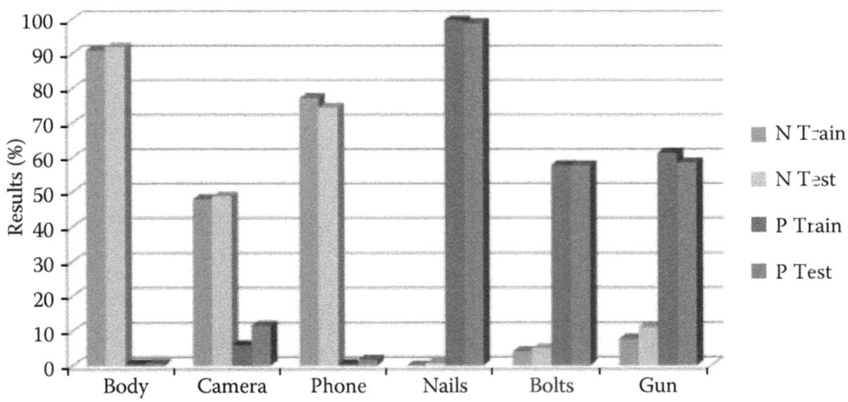

FIGURE 3.21 (See color insert.)
Sample results of neural network training using examples with all six object types. The height of the bars represents the percentage probability that the network assigns either N (negative threat) or P (positive threat), both for the set used in training the network and for the test set. (From Harmer, S. W. et al., *IEEE Sensors Journal*, 99, 2013, doi: 10.1109/JSEN.2013.2273487.)

at ranges of 4, 6, and 8 m included in the training set. The performance of the network can be assessed by how well it is able to classify the training data and, more importantly, how well it classifies a second, unseen, test set. Good agreement between the train and test results is important, as it shows that the network can generalize to new scenarios which haven't been previously encountered.

The output of the network is a value between 0 and 1, which can be interpreted as a confidence level that a threat has been encountered. A value close to 1 indicates a high confidence that a threat is present, and conversely, a value close to 0 indicates a high confidence that no threat is present. A value around 0.5 indicates that the network is unable to make a reliable decision. In practice, all results between 0.2 and 0.8 are discarded, which usually means that N and P results do not sum to 100%, as can be seen in Figure 3.21.

The results displayed in Figure 3.21 show that much improved discrimination is possible, not just between body and fragmentation devices, but also for more problematic cases of distinguishing a handgun from a camera. In this case, unlike the previous analysis using thresholds alone, a strong correct classification of these and other objects is now displayed in Figure 3.21, and with low false positive rates. This improvement comes from not only an optimal choice of thresholds, but also increased information from the depth spectrum, which gives information about the size and structure of the object in the range direction.

The effect of reducing the range of training objects is shown in Figures 3.22 and 3.23. In Figure 3.22 the most scattering object (nails) has been omitted from the training; as seen, deleting objects which scatter strongly has little

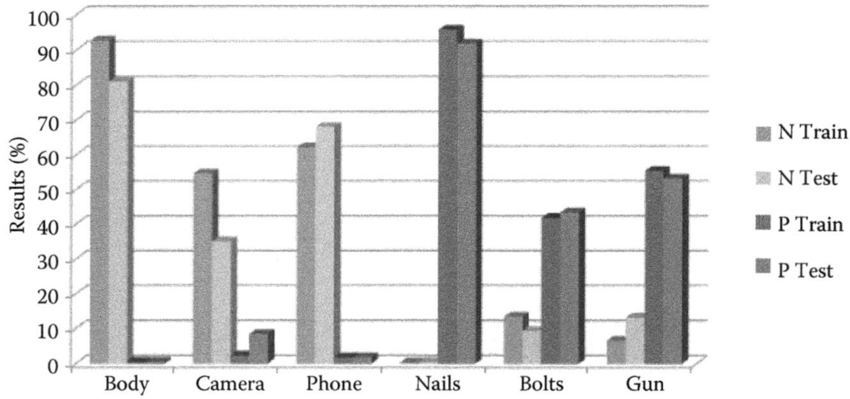

FIGURE 3.22 (See color insert.)
Examples of results from neural network training using a limited set of training objects. Results are shown as a percentage of classification results. (From Harmer, S. W. et al., *IEEE Sensors Journal*, 99, 2013, doi: 10.1109/JSEN.2013.2273487.)

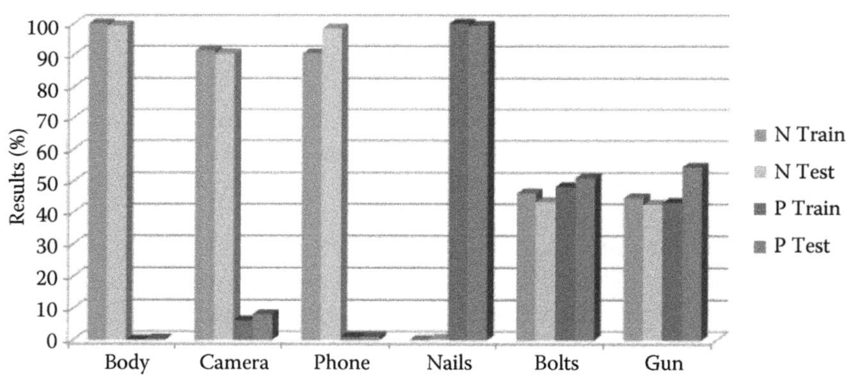

FIGURE 3.23 (See color insert.)
Examples of results from neural network training using only a PBIED with nails and the body alone in the training set. (From Harmer, S. W. et al., *IEEE Sensors Journal*, 99, 2013, doi: 10.1109/JSEN.2013.2273487.)

effect on the overall classification capability for all the objects. Concentrating on the difficult to distinguish objects has previously been shown to be advantageous in training [13] due to more efficient use of the ANN capabilities, though this does require the ability to extrapolate to more extreme cases when required.

Using extreme cases for training and attempting to classify similar objects is generally much less effective, as can be seen in Figure 3.23, where only the extreme items are included in the training; the results for the gun and smaller PBIED are much less certain. As shown in Figures 3.21–3.23, the network has the property of generalizing to other objects, though in practice as

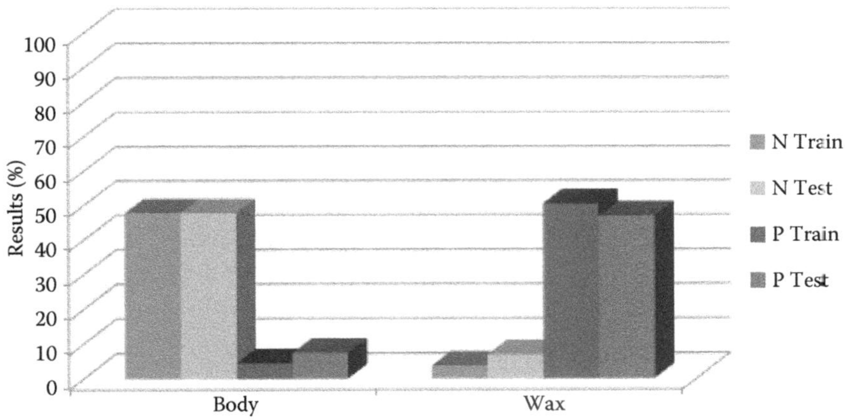

FIGURE 3.24 (See color insert.)
Sample results from neural network training, showing the discrimination of nonmetallic fragmentation PBIEDs. (From Harmer, S. W. et al., *IEEE Sensors Journal*, 99, 2013, doi: 10.1109/JSEN.2013.2273487.)

wide a range as possible of likely clutter objects should be included in the training set for best performance.

The network method is also capable of identifying the presence of dielectric-only targets, such as PBIEDs without fragmentation, by using the increased signal peak shown in the co-polar depth spectrum in Figure 3.11. A wax block strapped to the torso was used as a simulant, and the result of training is shown in Figure 3.24. The detection rate (~50%) is lower than for the fragmentation devices, due to the sensitivity to alignment with respect to the beam, but is much greater than the false positive rate (<5%). Multiple polling of results allows these rates to be improved further.

It is important to note that this method is mainly sensitive to the dielectric constant of the concealed material and its thickness [5], [31] and does not reveal further spectroscopic information as to the identity of the material. It is therefore not possible to distinguish, for example, plastic explosives from concealed material with similar dielectric properties. It should therefore be combined with additional visual clues from the video image and with the positioning of the suspect area on the torso. It may also be noted that explosive devices with a fragmentation component, including those made with dielectric materials only, also produce cross-polar signatures in a manner similar to that for metal materials.

The sensor operates well with moving targets, as walking people tend to present multiple aspects of any concealed items to the radar over time. However, when a person turns, such that his or her torso is highly oblique (~45° or more) to the sensor, the radar return in both polarization states can be significantly reduced; this is a great problem if threshold-only detection is used, but can be ameliorated to a considerable degree by the more sophisticated depth profile analysis discussed. The reasons for this are that the concealed object will often show a depth profile independent of angle, which allows detection in many aspects, even though much of the incident millimetre radiation is scattered away from the receiver horns.

The detection of PBIED with metallic fragments presents the most reliable and easily discriminated concealed object; smaller objects such as handguns are less easily distinguished from benign objects (such as a camera), but can still be detected and discriminated from these under good conditions where the user can interrogate the person from different directions, for example, by having the person being screened perform a full rotation.

3.5 Handheld Version of Sensor

A handheld, portable active millimetre-wave sensor, using a synchronized amplitude-modulated (AM) detector, is shown schematically in Figure 3.25 [17, 22].

It consists of transmitting (T_x) and receiving (R_x) units. The two identical W-band horn-lens antennae with 5° beamwidth and gain ~30 dBi are used

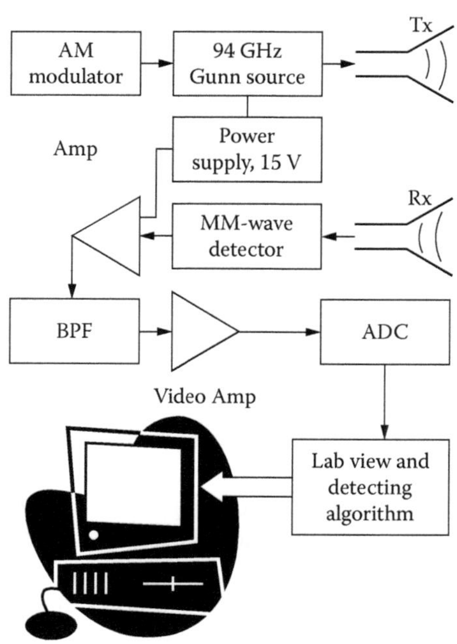

FIGURE 3.25
Simplified block diagram of the millimetre-wave sensor based on synchronized detection. (From Kapilevich, B. and Einat, M., *IEEE Sensors Journal*, 10(11), 1746–1752, 2010.) © 2010 by the Institute of Electrical and Electronics Engineers [IEEE].)

for transmission and reception. The CW Gunn diode generator produces an output power of 13 dBm at 94 GHz; this is modulated at 1 KHz, resulting in an AM signal at the generator's output. The backscattered signal is captured by the R_x antenna and delivered to a millimetre-wave, zero-biased Schottky detector with an input sensitivity of 800 mV/mW. The detector, utilizing MIC technology and Schottky barrier beam lead diodes, provides a very economical solution for power detection over the 18–330 GHz range. High-sensitivity and full-waveguide bandwidths are achieved simultaneously, without external DC bias or adjustments. The detector has a flat frequency response, as its sensitivity shows minor variation over the entire waveguide band. After amplification and filtering by a band pass filter with a 1 KHz centre frequency, the signal goes to the input of an analogue-to-digital converter (ADC). The ADC is integrated with a LabView interface which is matched with a detecting algorithm written using MATLAB® code.

The sensor is supplied by 15 V DC and its general view is shown in Figure 3.26. The W-band and low-frequency components of the sensor are commercially available at relatively low cost. The overall system's noise performance is characterized by a signal-to-noise ratio of 3 dB at a distance from the target of 3 m.

Two metallic rods with diameter D = 1.2 cm, positioned with a distance of 10 cm between them and located 50 cm from the sensor, have been tested (Figure 3.27a). The proper recorded signature is shown in Figure 3.27b for single forward and backward scans. One can see that a spatial resolution of 10 cm, or sometimes less, is achievable for the sensor considered.

As an explosive material, a simulant of C4 has been employed. It is quite a lossy media with a low real part of dielectric constant. When concealed on the human body, it absorbs a part of illuminating radio frequency (RF) power. As a result, the reflected signal is lower, in comparison with the same signal reflected

FIGURE 3.26
General view of W-band sensor.

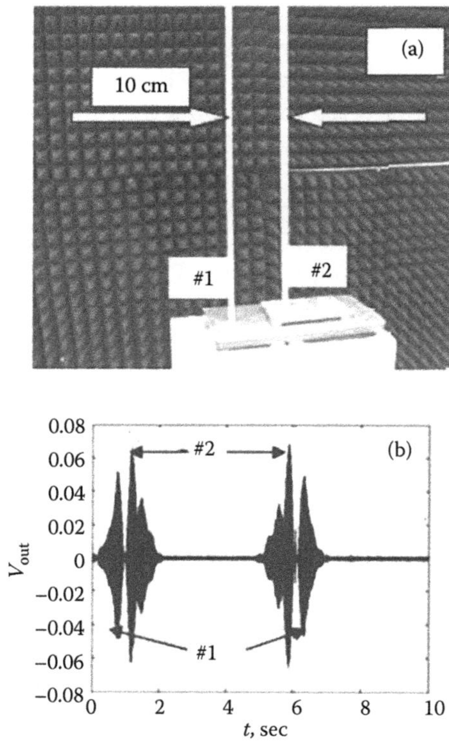

FIGURE 3.27
Experimental setup used for (a) estimating spatial resolution and (b) its signature. The distance between metallic rods is 10 cm. (From Kapilevich, B. and Einat, M., *IEEE Sensors Journal*, 10(11), 1746–1752, 2010. © 2010 by the Institute of Electrical and Electronics Engineers [IEEE].)

by a human body without explosive. This effect is illustrated by Figure 3.28. It should be noted that the recorded signatures are dependent on orientation of the object leading to change of the reflected amplitude. Figure 3.29 shows a person with explosive and gun participating in the experiments.

Careful analysis of results, with a variety of different targets, shows that when the antenna's beam crosses the boundary between a human body and the edge of a target, a sharp peak in response appears. This effect can be used to develop an algorithm which enhances the detection probability and reduces the false alarm rate. The time gating characteristic and reflected power threshold level must be taken into consideration, simultaneously, to arrive at a decision about presence or absence of any concealed objects. Figure 3.30 illustrates a typical output signal of the sensor after scanning an area containing 300 spatial sampling points. The person participating in the experiment is located within the range of 50–200 sampling points (Figure 3.30a). The same record for the gun hidden under shirt is depicted in Figure 3.30b. The distance between the tested person and the sensor was

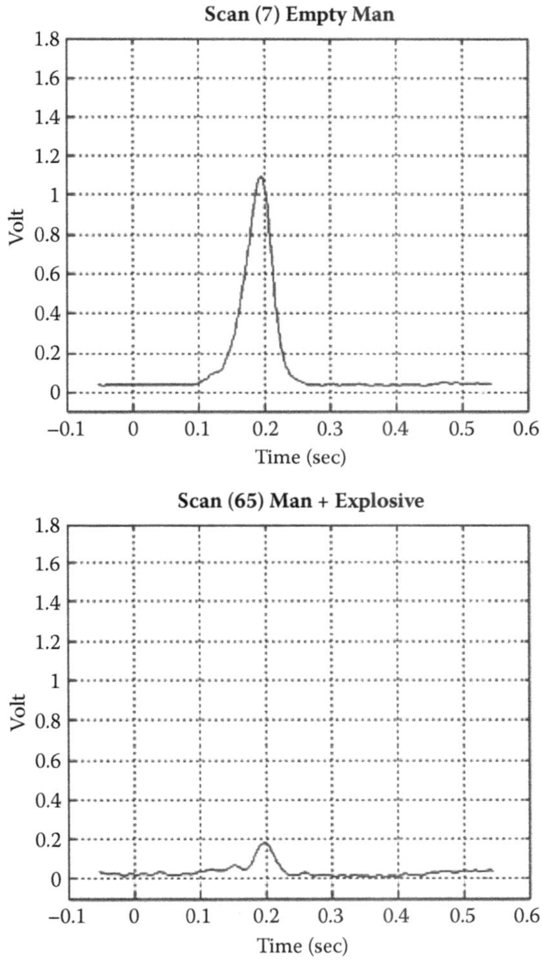

FIGURE 3.28
Comparison reflected signals from a human body without and with an explosive. (From Kapilevich, B. and Einat, M., *IEEE Sensors Journal*, 10(11), 1746–1752, 2010.) © 2010 by the Institute of Electrical and Electronics Engineers [IEEE].)

3 m. It should be pointed out that clutters such as parasitic reflections, from walls and other objects located in proximity to the person, may be a source of false alarm. To avoid the effect of such undesirable clutter, we have employed only a part of the scanning sector within 50–200 sampling intervals corresponding to the actually active scanning sector.

Comparing both signatures, we can observe the unique peak on Figure 3.30b, between 170–175 sampling points. The magnitude of the peak exceeds 0.15 V. Assuming that the given peak corresponds to the hidden object, we may enhance the detection procedure by adding the time gating threshold magnitude of the reflected power level in the processing algorithm,

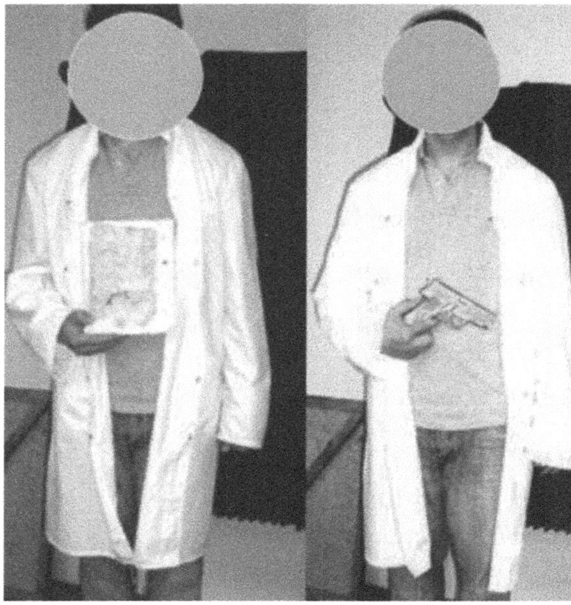

FIGURE 3.29
A person with a plastic explosive and gun participating in the experiments. (From Kapilevich, B. and Einat, M., *IEEE Sensors Journal*, 10(11), 1746–1752, 2010. © 2010 by the Institute of Electrical and Electronics Engineers [IEEE].)

namely, four sampling intervals in our case. Based on this criterion, the following detecting algorithm has been suggested:

1. Data are collected from a certain scan, and all the samples that are below the selected threshold are ignored as 'no detection'.
2. When a sample is above the threshold, a time gate of Δ samples is activated. If the numbers of samples that are above the threshold are Δ or less, then detection is declared 'true'.
3. Else, if the signal stays above the threshold for more than Δ samples, it is regarded as 'not a sharp' echo and is ignored.

The algorithm was applied to samples 50–200, which are the boundaries of the active checked area. Outside this range, the signals come from metal holders far outside the human body, which are disregarded by the algorithm. Sometimes, there is a parasitic reflection from an 'empty' human body (carrying no objects) satisfying the both predefined criteria resulting in a false alarm. Also, the reflection not satisfying these criteria leads again to a false alarm. Therefore, the total percentage of false alarm and true decisions may exceed 100%.

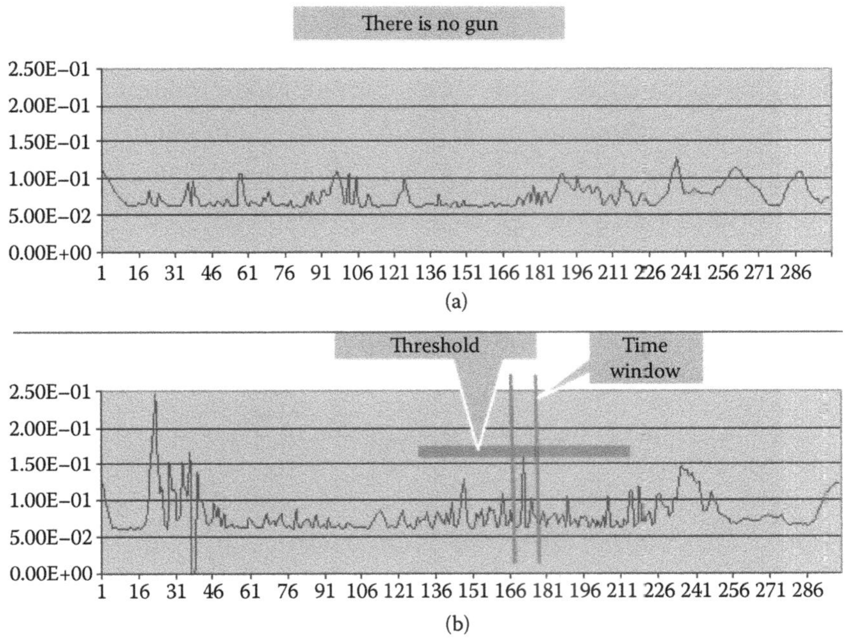

FIGURE 3.30
Comparison of reflected signals from a human body (a) without a gun and (b) with the hidden gun. (From Kapilevich, B. and Einat, M., *IEEE Sensors Journal*, 10(11), 1746–1752, 2010. © 2010 by the Institute of Electrical and Electronics Engineers [IEEE].)

Since the decision is based on a relative measurement, it is important to set a uniform reference level. This is done by performing each experimental session under the same conditions where the parameters are quite constant. Still, there may be a variation in these parameters. In order to avoid this, two objects were placed at the ends of the target area: a corner reflector and an absorbing material. At the start of the experiment, readings are taken from these objects and used as maximum and noise level, respectively; these provide calibration or reference levels for the system. During the experiment, the sensor is directed periodically at these objects and the levels are monitored. If a change is noticed, it is compensated for in the amplifying chain. Therefore, uniform reference levels are maintained.

When a static target (a barrel of water, for example) is measured, the readings obtained are reproducible. But, when measurements were taken on a human body, the readings were changing over different measurements of the same scenario. It was found that when a human body is measured, although the person tries to remain stationary, he or she moves. There were also minor variations due to uncontrolled movements, and the person still moves even when asked to hold his or her breath. These variations are on the

TABLE 3.1

Summary of the Measured Threshold Magnitudes and Time Gates for 10 Detection Trials of a Human Body with and without Hidden Gun

	Without Gun			With Gun	
No	Magnitude, V	Time Gate, Δ	No	Magnitude, V	Time Gate, Δ
1	0.14	1	11	0.156	1
2	0.1	2	12	0.172	2
3	0.096	5	13	0.112	2
4	0.108	2	14	0.144	1
5	0.154	2	15	0.166	2
6	0.148	6	16	0.172	1
7	0.114	1	17	0.174	1
8	0.106	3	18	0.168	1
9	0.168	5	19	0.18	1
10	0.12	3	20	0.208	4

TABLE 3.2

Percentages of True and False Decisions Estimated with the Threshold Magnitudes and Time Gate Criteria

Magnitude, V	Time gate, Δ	False %	True %
0.13	Not specified	40	90
0.14	Not specified	30	85
0.15	Not specified	20	80
0.13	1	10	90
0.14	1	5–10	95–100
0.15	1	20	100
0.13	2	20	90
0.14	2	10–20	95–100
0.15	2	10	100

millimetre-scale range and are detectable with the sensor. Therefore, a different reading is obtained each time.

In order to verify the above approach, 10 search trials of the human body without a weapon and 10 trials with a concealed gun were taken. The person was asked to remain as still as possible. Table 3.1 summarizes the measure threshold magnitudes and time gate Δ (sampling intervals). Using these data the true–false percentage has been estimated on the basis of the following settings: threshold magnitude, 0.13–0.15 V; sampling interval, 1 and 2 points or not specified. The percentages of true and false decisions estimated with the above criteria are given in Table 3.2. One can see that employing only backscattered magnitude without time gating leads to a high false alarm rate. However, adding time gating and minimizing the threshold magnitude gives a reduction in false alarm rate and an increase in detection rate. For example, in the case of magnitude V = 0.13 V and time gate $\Delta = 1$, we have obtained 10% (false) + 90% (true) = 100%!

3.6 Conclusion

An active millimetre-wave radar system has been described for detection of PBIEDs and handguns at standoff distances. Using a W-band, UWB radar comprised of commercial off-the-shelf components, the system has the capability to detect and classify concealed objects in real time, at video refresh rates at ranges of up to 25 m. A fast response time has been achieved through use of direct detector amplifiers and a fast data acquisition system. The prototype system is tripod mounted and manually steered and aimed by an operator who relies on video imagery to guide the millimetre-wave radar beam onto the person to be screened. Effective detection of simulated PBIED objects hidden under clothing at ranges up to 25 m makes the system a potentially valuable screening tool for scenarios where the threat of suicide attacks is considered possible. The identification of objects such as handguns and discrimination of these from other commonly carried clutter objects is possible, but at a reduced standoff range and with a lower probability of detection. The sensor described has several obvious scenarios where such a device would be useful: a pre-screening method prior to people arriving at a checkpoint, where more conventional screening techniques may then be applied; a vehicle-mounted device for discrimination of combatants from civilians; and as a primary screening tool for passengers passing through underground or metro-style ticket barriers. See Table 3.3 for typical performance in this scenario.

TABLE 3.3

Typical Results Obtained in a Blind Walking Test Carried Out over 6–20 m Standoff

Target	True Positive	True Negative	False Positive	False Negative
PBIED	84%	0	0	16%
Handgun	50%	0	0	50%
Unarmed	0	97%	3%	0

References

1. N. Levanon, *Radar principles*, John Wiley & Sons, New York, 1988, pp. 1–2.
2. D. A. Andrews, S. E. Smith, N. D. Rezgui, N. J. Bowring, M. Southgate, and S. W. Harmer, A swept millimetre-wave technique for the detection of concealed weapons and thin layers of dielectric material with or without fragmentation, *Proceedings of SPIE*, 7309, 2009.
3. D. A. Andrews, S. W. Harmer, N. J. Bowring, N. D. Rezgui, and M. J. Southgate, Active millimeter wave sensor for standoff concealed threat detection, presented at IEEE Sensors 2013.
4. M. Gashinova, V. Djigan, L. Y. Daniel, and M. Cherniakov, Adaptive calibration in UWB radar, in *Proceedings of IEEE Radar Conference*, May 2008, pp. 1161–1166.
5. D. A. Andrews, N. Rezgui, S. E. Smith, N. J. Bowring, M. Southgate, and J. G. Baker, Detection of concealed explosives at standoff distances using wide band swept millimetre waves, *Proceedings of SPIE*, 7117, 2008.
6. N. J. Bowring, J. G. Baker, N. Rezgui, M. Southgate, and J. F. Alder, Active millimetre wave detection of concealed layers of dielectric material, *Proceedings of SPIE*, 6540, 2007.
7. N. J. Bowring, J. G. Baker, N. D. Rezgui, and J. F. Alder, A sensor for the detection and measurement of thin dielectric layers using reflection of frequency scanned millimetric waves, *Measurement Science and Technology*, 19(2), 024004, 2008.
8. N. J. Bowring, J. G. Baker, and J. F. Alder, Detection and measurement of thin dielectric layers using reflection of scanned millimetric waves, in *Proceedings of IEEE International Conference on Networking, Sensing and Control*, April 2007, pp. 437–442.
9. D. A. Andrews, N. J. Bowring, M. Southgate, E. Guest, S. W. Harmer, and A. Atiah, A multifaceted active swept millimetre-wave approach to the detection of concealed weapons, *Proceedings of SPIE*, 7117, 2008.
10. A. Agurto, Y. Li, G. Y. Tian, N. Bowring, and S. Lockwood, A review of concealed weapon detection and research on perspective, *Proceedings of IEEE ICNSC*, 443–448, 2007.
11. N. Bowring, D. Andrews, N. D. Rezgui, and S. W. Harmer, Remote detection and measurement of objects, UK Patent 2,458,764, 18 March 2009.
12. S. W. Harmer, N. J. Bowring, D. Andrews, and N. D. Rezgui, A review of non-imaging stand-off concealed threat detection with millimeter-wave radar, *IEEE Microwave Magazine*, 13(1), 160–167, 2012.
13. J. W. Lamb, Miscellaneous data on materials for millimetre and submillimetre optics, *International Journal of Infrared Millimeter Waves*, 17(12), 1997–2034, 1996.
14. R. Buchner, G. T. Hefter, and P. M. May, Dielectric relaxation of aqueous NaCl solutions, *Journal of Physical Chemistry A*, 103(1), 1–9, 1998.
15. N. Yamada, Radar cross section for pedestrian in 76 GHz band, *R&D Review of Toyota CRDL*, 39(34), 2004.
16. R. J. Douglass, J. D. Gorman, and T. J. Burns, System and method for standoff detection of human carried explosives, U.S. Patent 7,800,527, 11 October 2005.
17. B. Kapilevich and M. Einat, Detecting hidden objects on human body using active millimeter wave sensor, *IEEE Sensor Journal*, 10(11), 1746–1752, 2010.

18. J. Hausner and N. West, Radar based concealed threat detector, in *IEEE MTT-S International Microwave Symposium*, 2007, pp. 765–768.
19. J. Hausner and J. M. West, Object detection method and apparatus, U.S. Patent 0052576, 8 March 2007.
20. Available at http://arielrnd.com/.
21. Available at http://www.detectanddefeat.com/.
22. B. Kapilevich and M. Einat Hand-held device and method for detecting concealed weapons and hidden objects. US2009195435(A1)-2009-08-06.
23. Available at http://www.setcorp.com/.
24. R. Appleby and R. N. Anderton, Millimeter-wave and submillimeter-wave imaging for security surveillance, *Proceedings of IEEE*, 95. 1683–1690, 2007.
25. R. Appleby and H. B. Wallace, Standoff detection of weapons and contraband in the 100 GHz to 1 THz region, *IEEE Transactions on Antennas and Propagation*, 55(11), 2944–2956, 2007.
26. S. Oka, H. Togo, N. Kukutsu, and T. Nagatsuma, Latest trends in millimeter-wave imaging technology, *Progress in Electromagnetic Research Letters*, 1, 197–204, 2008.
27. B. Kapilevich, Y. Pinhasi, B. Litvak, M. Anisimov, and D. Hardon, Complex permittivity measurements of textiles and leather in a free-space: an angular-invariant approach, *International Journal of Microwave Science and Technology*, 2012, 375601, 2012. Available at http://www.hindawi.com/journals/ijmst/2012/375601/.
28. K. B. Cooper, R. J. Dengler, N. Llombart, T. Bryllert, G. Chattopadhyay, E. Schlecht, J. Gill, C. Lee, A. Skalare, I. Mehdi, and, P. H. Siegel, Penetrating 3-D imaging at 4- and 25-m range using a submillimeter-wave radar, *IEEE Transactions on Microwave Theory Technology*, 56(12), 2771–2778, 2008.
29. N. A. Salmon, P. Wilkinson, and C. Taylor, Interferometric aperture synthesis for next generation passive millimetre wave imagers, presented at SPIE Europe Security + Defence, Millimetre Wave and Terahertz and Technology, Edinburgh, September 2012.
30. S. A. Shilo and Yu. B. Sidorenko, Millimeter wave imaging system, in *Sixth International Kharkov Symposium on Proceedings of Physics and Engineering of Microwaves, Millimeter and Submillimeter Waves and Workshop on Terahertz Technologies (MSMW '07)*, vol. 1, 2007.
31. S. W. Harmer, N. J. Bowring, N. D. Rezgui, and D. Andrews, A comparison of ultra wide band conventional and direct detection radar for concealed human carried explosives detection, *Progress in Electromagnetics Research Letters*, 39, 37–47, 2013.
32. N. Rezgui, D. Andrews, N. Bowring, S. Harmer, and M. Southgate, Standoff detection of concealed handguns, *Proceedings of SPIE*, 6948, 2008.
33. L. Yong, Y. T. Gui, N. J. Bowring, and N. Rezgui, A microwave measurement system for metallic object detection using swept frequency radar, *Proceedings of SPIE*, 7117, 2008.
34. S. W. Harmer, D. Andrews, N. J. Bowring, N. D. Rezgui, and M. Southgate, Active millimeter wave sensor for standoff concealed threat detection, *IEEE Sensors Journal*, 99, 2013. DOI: 10.1109/JSEN.2013.2273487.

4

FMCW Sensors for Detecting Hidden Objects

Frequency-modulated continuous wave (FMCW) signals are widely used in radar technologies currently deployed for wide area surveillance, site security, anti-collision transport systems and level control sensors, including altimetres and imaging systems. Sensors using FMCW also can be designed for detecting concealed objects. This chapter focuses on various aspects of FMCW which relate to the detection of concealed objects.

4.1 Linear FMCW Theory

FMCW relies on the difference between transmitted and received frequency of a frequency-swept waveform. The linearly changing instantaneous frequency f_i and phase φ_i of the carrier wave over the sweep period can be presented [1], [2] as

$$f_i = f_c + \frac{\Delta f}{T_{sweep}} \cdot t \text{ and } \varphi_i(t) = 2\pi \int f_i(t)dt = 2\pi\left(f_c t + \frac{\Delta f}{2T_{sweep}} t^2 \right) \quad (4.1)$$

where the ratio $\Delta f / T_{sweep}$ is defined as the compression ratio in Hz/s and f_c is the carrier frequency. As a result, a modulated linear FM signal, $y_{Linear\text{-}FM}(t)$, is obtained in the following form:

$$y_{Linear-FM}(t) = A_c \cos\left[2\pi f_c t + \pi \frac{\Delta f}{T_{sweep}} t^2 + \phi_{VCO}(t) \right] \quad (4.2)$$

where $\phi_{VCO}(t)$ is the phase noise generated by the oscillator.

The signal reflected from a target is time delayed, depending on the distance d, so that the round-trip time delay τ is

$$\tau = \frac{2d}{c} \quad (4.3)$$

where c is the velocity of light: $c \cong 3 \cdot 10^8$ m/s. The resulting signal at the receiver's input is now a time-delayed version of Equation 4.2.

$$y_{Linear-FM}(t-\tau) = A\cos\left[2\pi f_c(t-\tau) + \pi\frac{\Delta f}{T_{sweep}}(t-\tau)^2 + \theta(t) + \phi_{VCO}(t-\tau)\right] \quad (4.4)$$

where $\theta(t)$ is the instantaneous phase change in time of the FM signal.

Figure 4.1 illustrates the basic operational concept of FMCW sensors. A linear, frequency-modulated signal, also named a chirp, is transmitted in a certain direction and scattered from the target. The delayed echoes are captured by the receiving antenna and mixed with the transmitted signal, resulting in an intermediate frequency (IF) signal. The signal is then amplified by a video amplifier and filtered with a low-pass filter to suppress the high-order frequency products of the mixer.

Figure 4.2 presents the transmitted and return signals from a target when the period is T and the frequency of the FMCW signal varies linearly over a range of Δf. The time delay between transmission and reception depends on

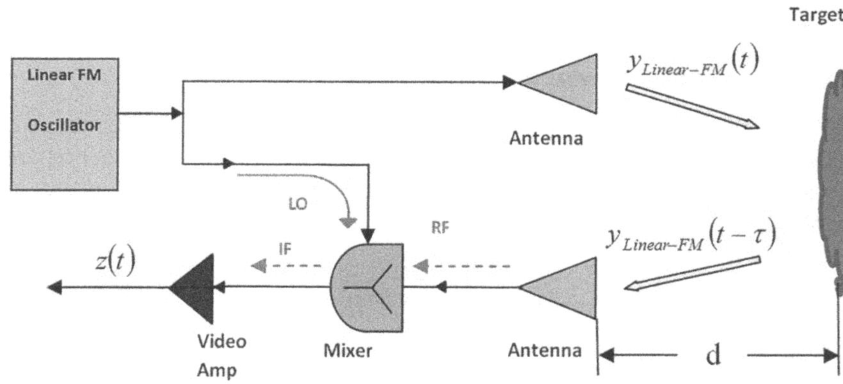

FIGURE 4.1
Illustration of the basic operational concept of FMCW sensors.

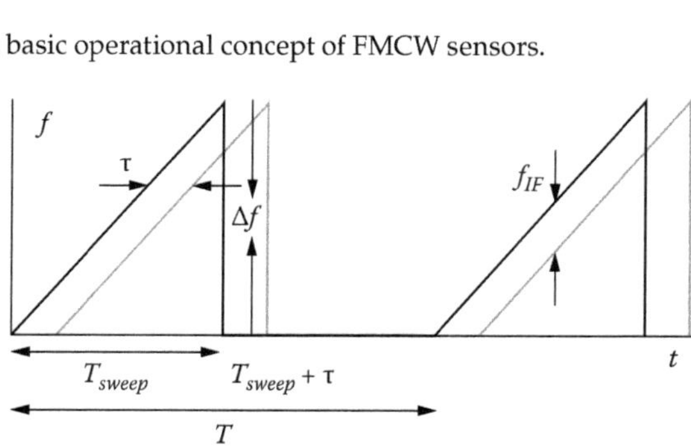

FIGURE 4.2
The instantaneous frequency f of the linear FM transmitted and received signals in the FMCW sensors.

the distance *d*. The IF signal after mixing and filtering at the receiver output is given by

$$z(t) = \frac{A \cdot A_c}{2} \cos\left[\underbrace{2\pi \cdot \underbrace{\frac{\Delta f}{T}}_{f_{IF}} \cdot \tau \cdot t + 2\pi f_c \tau - \pi \frac{\Delta f}{T} \tau^2 - \theta(t)}_{\varphi(t)} + \underbrace{\phi_{VCO}(t) - \phi_{VCO}(t-\tau)}_{\text{Phase noise}} \right] \tag{4.5}$$

By keeping a constant sweep time, the intermediate frequency f_{IF} is proportional to the distance and can be easily measured.

The spectra of output signal can be estimated as follows [3]:

$$\nu_{IF} = \frac{1}{2\pi}\frac{d\varphi(t)}{dt} = \underbrace{\frac{\Delta f}{T_{sweep}}}_{f_{IF}} \cdot \tau - \underbrace{\frac{1}{2\pi}\frac{d\theta(t)}{dt}}_{\text{Doppler}} + \underbrace{\frac{1}{2\pi}\frac{d}{dt}\left[\theta_{VCO}(t) - \theta_{VCO}(t-\tau)\right]}_{\text{FM noise}} \tag{4.6}$$

It contains, in general, the FM noise, Doppler, and intermediate frequency components. The latter is proportional to the distance from the target detected:

$$f_{IF} = \frac{\Delta f}{T_{sweep}} \cdot \tau = 2\frac{\Delta f}{T_{sweep}} \cdot \frac{d}{c} \tag{4.7}$$

Using a Fourier transformation the spectrum of beat frequencies is analyzed to acquire information about the target—the backscattering amplitude and the distance to the target. However, different factors leading to an inaccuracy in determination of the intermediate frequency are responsible for finite range resolution. Assuming that the actual frequency resolution is δf_{IF}, the corresponding range resolution δd is

$$\delta d = \frac{cT_{sweep}}{2\Delta f}\delta f_{IF} \tag{4.8}$$

It can be shown [29] that $\delta f_{IF} \approx T_{sweep}$, resulting in a simple relationship between the bandwidth of a chirp and the range resolution:

$$\Delta d = c/2\Delta f \tag{4.9}$$

Hence, it follows from Equation 4.9 that for better range resolution the chirp bandwidth should also be increased. However, some constraints such as limited operating bandwidth of R_x and T_x channels, non-linearity of the chirp pulse and leakage and phase noise effects must also be taken into account in estimating the true range resolution of the designed sensors. There are several approaches suggested to improve the range resolution [29]:

- Correction of the VCO linearity using a lookup table
- Correction of the VCO temperature drift
- Implementing surface acoustic wave delay lines
- Using the direct digital synthesis with an up-conversion stage

4.2 Basic Hardware Configurations

FMCW sensors can be classified into two categories:

- Sensors with the single antenna for both R_x and T_x operations—monostatic configuration that is typical for a radar
- Sensors with the separated antennae for both R_x and T_x operations—bistatic configuration

However, in many practical realizations the R_x and T_x antennae are placed nearby so that the subtended angle between receiver, target and transmitter is close to zero. Therefore, the second category can be considered as pseudo-monostatic.

Each category has inherent advantages and disadvantages. Principal advantages of the monostatic configuration are compactness and automatic collimation of both T_x and R_x beams, while the disadvantage is a low isolation between transmitting and receiving channels resulting in a significant leakage. Additional measures must be undertaken to reduce or suppress undesirable leakage, and this leads to complexity of a sensor and increasing cost. Leakage is responsible for degrading overall noise performance and limits the level of transmitted power depending on saturation of the front-end low-noise amplifier (LNA). The principal advantage of the bistatic configuration is a higher isolation between T_x and R_x channels, while the disadvantages are larger dimensions compared with the monostatic counterpart and necessity of accurate *alignment* of the directions of the receiver and transmitter antenna. This last point becomes very important for highly directive antennae.

4.2.1 FMCW Sensors with a Single Antenna

The simplified block diagram of the FMCW sensor with a single antenna is shown in Figure 4.3. The FMCW source generates a linear frequency chirp Δf during a time T_{sweep}. The energy of this chirp passes to the circulator's port

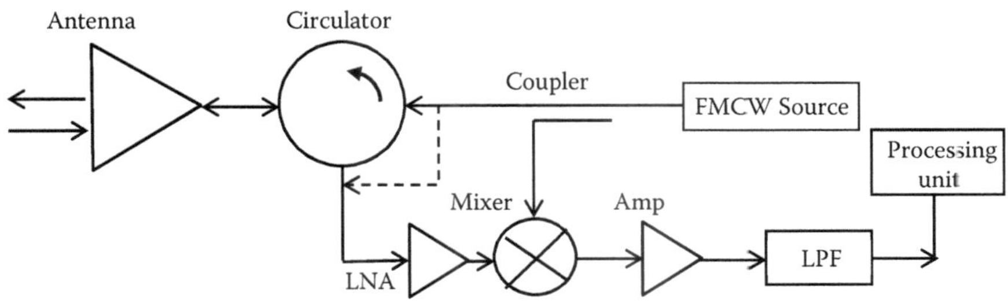

FIGURE 4.3
The simplified block diagram of the FMCW sensor with a single antenna.

that is directly coupled with antenna. The receiving echo is captured by the same antenna and goes to the circulator's port which is directly coupled with the radio frequency (RF) input port of the mixer. The local oscillator (LO) port of the mixer is driven by a portion of the radiated signal via directional coupler, resulting in the constant IF beat frequency at the mixer's output. Other spectral products are filtered by a low pass filter (LPF), and finally, the IF signal is analyzed by the processing unit, which is capable of performing a fast Fourier transform (FFT) process.

Because of a finite isolation of the circulator, a portion of the transmitted signal penetrates to the RF port of the mixer as shown by the dotted line in Figure 4.3. Assuming that the transmitted power is 10 dBm and the received signal power is –80 dBm or less, the overall dynamic range is 90 dB or more. Typical millimetre (mm)-wave Y-junction ferrite circulators have an isolation of about 20 dB. So, the leakage signal is about 70 dB higher than the received one. The overall performance of the sensor is degraded if the magnitudes of the echo response and the leakage signal are comparable. The latter is typical for weak backscattered signals from low radar cross section (RCS) targets at long operating distances. The magnitude and phase of the echo signal are functions of the target's range, while the similar parameters of the leakage are constant. Undesirable superposing (interference) of both signals reduces the sensitivity of the sensor, limiting operating range. The latter limits the level of transmitted power and requires improved antenna matching to minimize undesirable reflectance producing additional contributions to the leakage effects.

There are several approaches suggested to overcome the negative effect of leakage. For example, in the front-end FMCW radar module [13] the additional phase shifters were added in order to adjust the out-of-phase balancing between the leaked and internal reflections caused by non-ideal antenna matching. As a result, the minimum detectable signal can be improved by 15–20 dB. The block diagram shown in Figure 4.4 illustrates this approach. It should also be pointed out that if the phase shifters are frequency sensitive, the frequency chirp bandwidth, Δf, must be reduced, and that leads to degrading a resolution according to Equation 4.9.

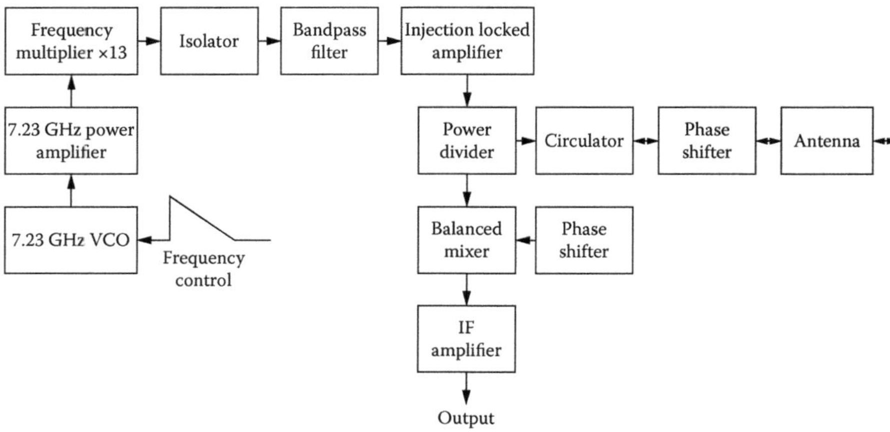

FIGURE 4.4
The front-end FMCW radar module with the additional phase shifters.

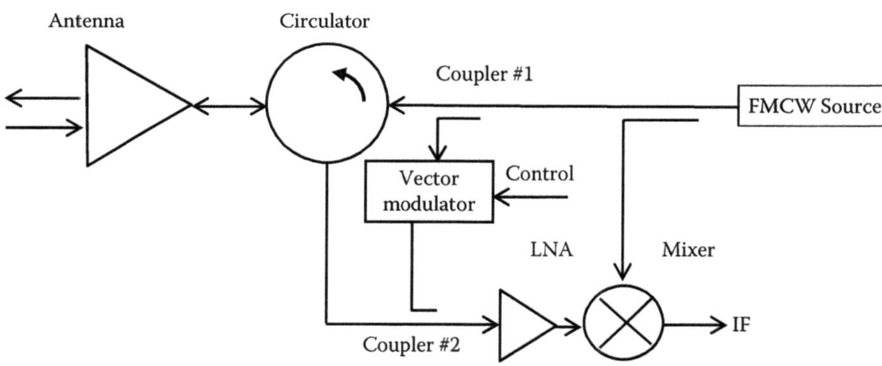

FIGURE 4.5
The simplified block diagram of the FMCW sensor with the leakage canceller.

Low isolation between the receiver and the transmitter prevents increasing transmitted power, as this may cause a saturation of the LNA front end. An efficient canceller of the leakage carrier was reported in [17]. A simplified block diagram of this device is depicted in Figure 4.5. The transmitted signal is sampled by directional coupler #1. The magnitude and phase of this signal are controlled by the vector modulator being the key element of the canceller. Using directional coupler #2 the sampled signal is injected to the R_x channel after the circulator. The vector modulator is adjusted to provide an appropriate magnitude and phase of the sampled signal needed to produce out-of-phase cancelling between the leakage and injected signals. If the out-of-phase condition is accurately controlled, an additional 30 dB of leakage suppression is possible. The complexity of the suggested cancelling circuitry

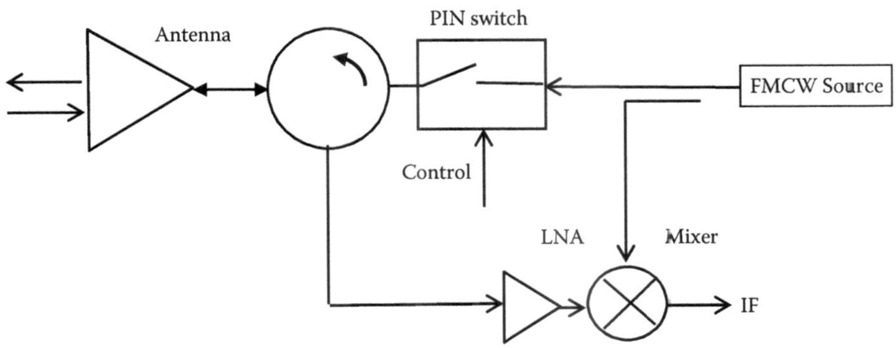

FIGURE 4.6
The simplified block diagram of the FMICW sensor with PIN switch.

limits implementation in inexpensive mm-wave sensors for detecting concealed objects.

The transmitting and receiving modes in the above versions of FMCW sensors operate simultaneously, and an increase of power automatically reduces the overall system performance due to destructive leakage. The frequency-modulated interrupted continuous wave (FMICW) mode employs the separate receive and transmit cycles as shown in Figure 4.6. When the PIN switch is ON the sensor operates in the conventional FMCW mode, but in the OFF state the transmitting cycle is interrupted while the receiving cycle is unchanged, permitting the mixer to obtain the IF at its output. As a result, isolation between R_x and T_x channels is essentially improved. Assuming a 50% duty factor, the average transmitted power is decreased by 3 dB in comparison with the unswitched transmit mode. However, due to better isolation the system noise performance is improved by more than 3 dB [18]. The latter depends on actual isolation of the PIN switch that also limits operational range caused by the transit time of the pin diode. An example of the W-band FMICW radar can be found in [19].

4.2.2 FMCW Sensors with Separated Antennae

A typical block diagram of such a sensor is shown in Figure 4.1. However, mutual coupling between antennae can also be a source of the parasitic leakage that depends on polarization and distance between antennae. Transmittance magnitude S_{21} between the two ports of identical horn antennae with radiating apertures 18 × 25 mm^2 and zero spacing has been simulated within the range 90–100 GHz for E- and H-planes using 3D electromagnetic (EM) solver (CST-12). Figure 4.7 shows the role of polarization in mutual coupling the two horn antennae. The comparison in both cases clearly demonstrates that the H-plane structure permits reducing the coupling level up to 80 dB due to a minimum electric field on the metal walls [35]. On the other hand,

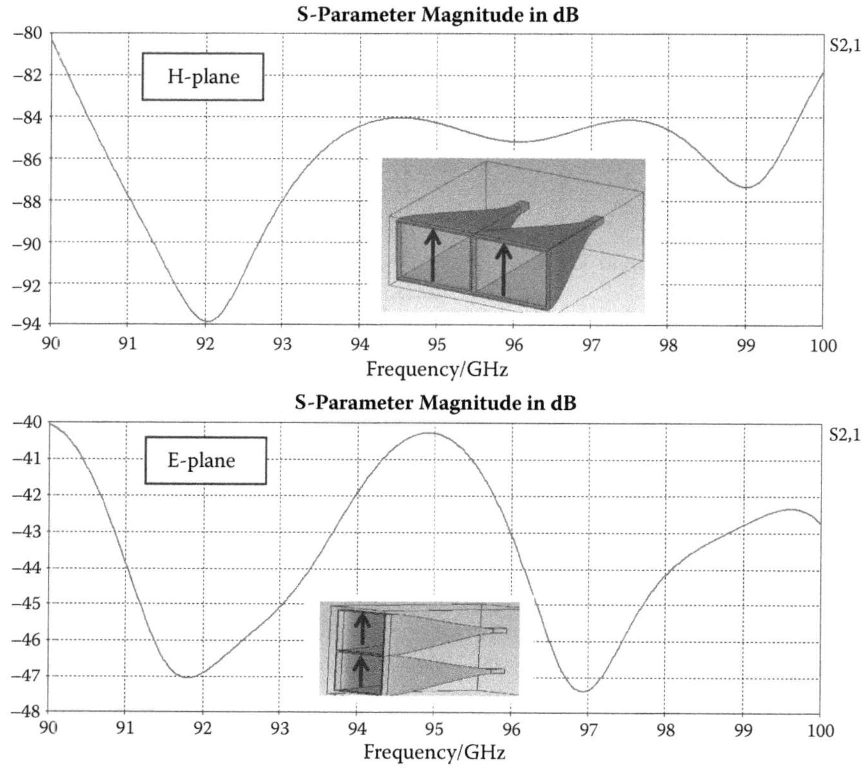

FIGURE 4.7
Transmittance magnitude S_{21} between the two ports for H- and E-plane configurations. (From Etinger, A. et al., *IEEE Sensors Journal*, 14, 4, 1232–1237, 2014. © 2010 by the Institute of Electrical and Electronics Engineers [IEEE].)

the E-plane structure suffers from edge diffraction, leading to an increase of mutual coupling.

From the systems point of view a sensor based on separated antennae is simpler than its single antenna counterpart since a circulator, PIN switch and additional couplers for leakage canceller are not required. However, the overall dimensions of a sensor with separated antennae preclude easily portable devices.

The contribution of the VCO phase noise to output the IF spectra is determined by the third term of Equation 4.6 regardless of whether a single antenna or two separated antennae are used. It can be shown [29] that in the presence of the leakage it is necessary to re-scale transmitted power P_t by subtracting the leakage in order to obtain the actual noise power density. As a result, the noise performance of the FMCW sensor is limited by the existing level of phase noise. The conventional way of improving the signal-to-noise ratio by increasing transmitted power does not hold true here due to the simultaneous increase in the noise floor, preventing improvement in system noise performance. Additional information about reducing the phase noise effect in FMCW radars can be found in [30]–[32].

4.3 W-Band Millimetre-Wave Sensors

During the last two decades millimetre-wave imaging systems operating in FMCW mode have been intensively developed for concealed weapon detection in various homeland security scenarios. However, such systems are expensive and sometimes can be replaced by simpler, non-imaging remote sensors detecting the suspected metal and plastic objects without reconstructing their shapes.

Non-imaging sensors operating with the FMCW mode determine the location of a hidden object with high accuracy. The block diagram of such a sensor operating in the W-band (94 GHz) is shown in Figure 4.8. [4]. A frequency synthesizer, HP-8350B, is the source of the FMCW signal in the range 15–16 GHz. This signal is divided and amplified by two power amplifiers (PAs). One of the PA drives an active 6× multiplier (model AMCL-10) generating a signal at W-band frequencies which is radiated by the T_x antenna. The second PA drives the same multiplier that is used in the R_x channel as the local oscillator of a balanced mixer. The signal scattered by a target is captured by the R_x antenna and, after amplifying by an LNA with a gain of 30 dB, comes to the RF input of the mixer. Both R_x and T_x channels employ the same

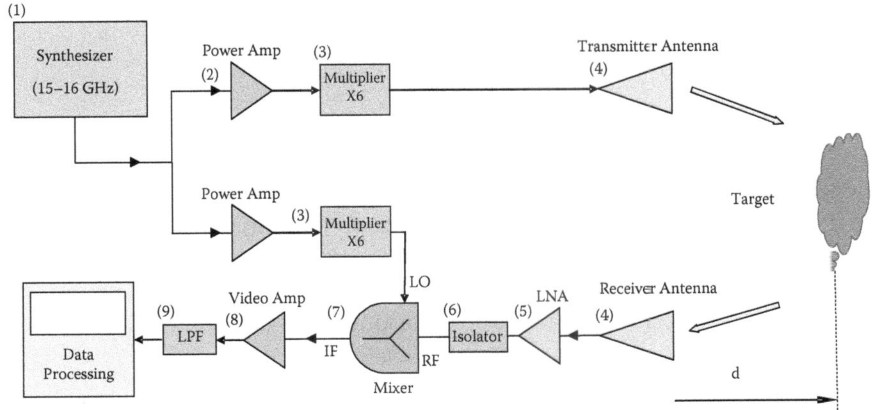

FIGURE 4.8
Block diagram of the W-band FMCW sensor: 94 GHz.

TABLE 4.1
W-Band Millimetre-Wave Sensor: Basic Components List

Number	Component
1	Frequency synthesizer HP 8530 B
2	Amplifier, Queen Star, QPJ-06183040-AO
3	Multiplier, Millitech, model AML-10-RFHOO
4	Horn-lens antenna, ELVA, model HLA-A-10/30
5	Low-noise amplifier, Queen Star, model QLN-940 46028-12
6	Isolator, Queen Star, model QIF WR-10
7	Balanced mixer, Elva, model BM-10-F
8	Video amplifier, Frequency Devices, model D-86
9	Low-pass filter 30 KHz

FIGURE 4.9
General view of the laboratory version of the W-band sensor.

horn-lens antennae (30 dBi gain and 5° beamwidth). The IF is proportional to the distance from a target, while its amplitude depends on scattering properties of a target and falls with a distance according to the expected inverse square relationship. A low-noise video amplifier is part of the processing unit. Table 4.1 summarizes the basic components of the sensor described. A general view of the sensor is shown in Figure 4.9.

4.3.1 Estimating System Performance

The received power can be estimated from the radar equation. Assuming a target with RCS = 10 cm^2 and a transmit power of P_t = 10 mW, the received power $P_{r,min}$ at a distance of 20 m can be calculated at frequency 94 GHz as follows:

$$P_{r,\min} = \frac{P_t \cdot G_r \cdot G_t \cdot \lambda^2 \cdot RCS}{(4\pi)^3 \cdot d^4}$$

$$= \frac{10 \cdot 10^{-3} \cdot 1000^2 \cdot (3.2 \cdot 10^{-3})^2 \cdot 10 \cdot 10^{-4}}{(4\pi)^3 \cdot 20^4}$$

$$= 3.225 \cdot 10^{-13} W$$

Suppose that other components are characterized by the following parameters:

- The LNA has a noise figure $F_{amp\_\max} = 6$ dB (4) and gain $G_{input} = 30$ dB (10^3).
- The insertion loss of the ferrite isolator $F_{isol} = 3$ dB (2) and gain $G_{isol} =$ –3 dB (0.5).
- The conversion loss of a mixer is $CL_{mix} = 7.8$ dB (6.026) and equivalent gain $G_{mix} = -7.8$ dB (0.166).
- The video amp has $F_{video} = 2$ dB (1.595) and gain $G_{video} = 20$ dB (10^2).
- The insertion loss of LPF is $F_{LPF} = 0.5$ dB (1.122) and gain $G_{LPF} =$ –0.5 dB (0.891).

The total system noise figure is found to be

$$F_{TOT} = F_{amp\_\max} + \frac{F_{isol} - 1}{G_{input}} + \frac{CL_{mix} - 1}{G_{input} G_{isol}} + \frac{F_{video} - 1}{G_{input} G_{isol} G_{mix}} + \frac{F_{LPF} - 1}{G_{input} G_{isol} G_{mix} G_{video}} .... =$$

$$= 4.02 \cong 6.04 \text{ dB}$$

This corresponds to the equivalent temperature $T_{eqv} = 290(F_{TOT} - 1) = 876$ K, assuming that the environment temperature is 290K..

Assuming that the bandwidth of the filter is BW = 30 KHz, the signal-to-noise ratio can be calculated as

$$\text{SNR} = 10 \log\left(\frac{\text{P}_{\text{r,min}}}{\text{k}_\text{B} \text{T}_{\text{eqv}} BW}\right) = 10 \log\left(\frac{3.225 \cdot 10^{-13}}{1.38 \cdot 10^{-23} \cdot 876 \cdot 3 \cdot 10^4}\right) = 10 \log(2630)$$

$$\cong 34.2 \text{ dB}$$

The noise floor is the noise power contributed by white noise at the equivalent temperature. The noise floor is found to be

$$NF = k_B T_{eqv} BW = 1.38 \cdot 10^{-23} \cdot 876 \cdot 3 \cdot 10^4 = -124.4 \text{ dBm}$$

The minimum detectable signal (MDS) of the receiver is the lowest power level at which it can detect an incoming signal in a presence of noise. This is an important parameter required for evaluation of the dynamic range of the receiver. So that

$$MDS = k_B \cdot T_{eqv} \cdot BW \cdot SNR$$

or in dBm,

$$MDS = -124.4 + 34.2 = -90.2 \text{ dBm}$$

Actual contributions of different RF components to the overall system performance depend on the configuration of hardware and may vary widely. For example, in a case of a single antenna, contribution of the leakage in degrading noise performance must also be taken into account.

4.3.2 Data Acquisition and Signal Processing

In order to record and sample the measured data a LabView interface has been developed. It includes the two channels: channel 1 records the output signal from the IF port, and channel 2 is used for recording the sawtooth modulation waveform and selecting the sampling rate, the number of samples and the file names to be recorded. The processing program is based on specially written MATLAB® code which forms the graphic user interface (GUI), as shown in Figure 4.10.

It permits reading the two files recorded by LabView: the reference sawtooth signal and the IF signal. It also processes the data by selecting an accurate time window of the IF signal with respect to the linear sweep times of the reference signal as depicted in Figure 4.11. When a target is scanned,

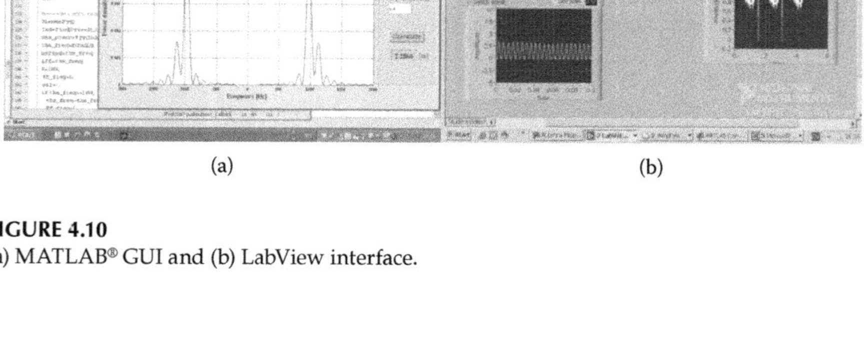

(a) (b)

FIGURE 4.10
(a) MATLAB® GUI and (b) LabView interface.

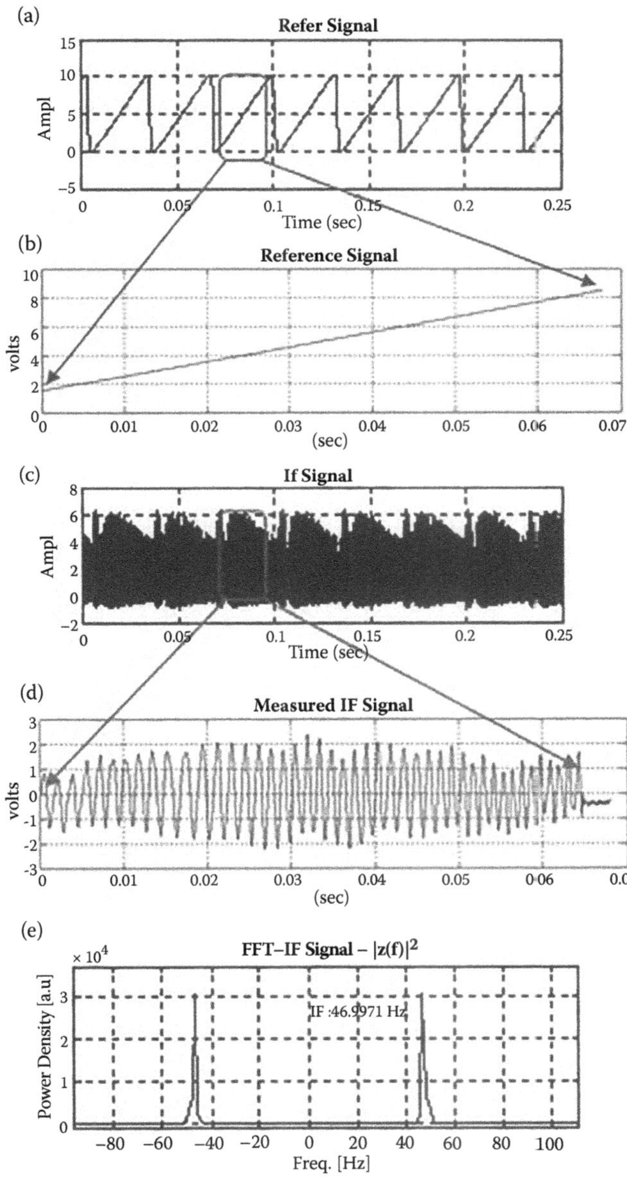

FIGURE 4.11
Illustration of the processing steps: (a, b) referenced signal of the linear modulation, (c, d) measured IF output, (e) result of application FFT procedure.

those windows are combined into one synchronized IF signal and, using a fast Fourier transform (FFT), calculate the IF frequency corresponding to the measured distance. Drawing the data with or without a DC component is optional. The window displayed at the right side shows the distance to the target resulting from this process.

4.3.3 Range Experiments

The estimation of the role of non-linearity in a frequency chirp on FMCW performance is an important consideration. The generated chirp signal is characterized by a central frequency of 15.346 GHz, a frequency span (Δf) of 200 MHz and a sweep time (T_{sweep}) of 0.25 s. Multiplication by a factor of 6 yields a 92.07 GHz signal with 1.2 GHz frequency span. The received IF data were sampled by using a 12-bit data acquisition card (DAQ) with a sampling rate of 22 KHz. Distance measurements were performed every 10 cm at a distance range of 5–10 m. The distance to the target is found as follows:

$$d = \frac{T_{sweep} \cdot f_{IF} \cdot c}{2 \cdot \Delta f} \tag{4.10}$$

By applying the FFT procedure with $N_f = 2^{14}$ samples, at the sampling rate f_s of 22 KHz, the resolution of the discrete Fourier transformation can be found:

$$\Delta f_{smpl} = \frac{f_s}{N_f} \tag{4.11}$$

The distance resolution accuracy has been determined using the expression

$$\Delta d = \left(\frac{\Delta f_{smpl} \cdot c}{2 \cdot {\Delta f}/{T_{sweep}}} \right) \tag{4.12}$$

and found to be 4.1 cm, where $\Delta f_{smpl} = 1.34$.

In case of linear FMCW, the distance calculated from the measured IF must be a linear function of the intermediate frequency. Figure 4.12 illustrates this statement; a good linearity of frequency sweeping is needed to reach range accuracy of several centimetres or less at a distance of 10 m. The results are consistent with the above theoretical prediction. Outdoor experiments have demonstrated that a gun covered by fabric can be easily detected at a distance of 10 m and more (Figure 4.13).

Indoor experiments were performed with three plastic sticks of diameter about 2 cm with a spacing of 15 cm at a distance of about 3 m from the sensor

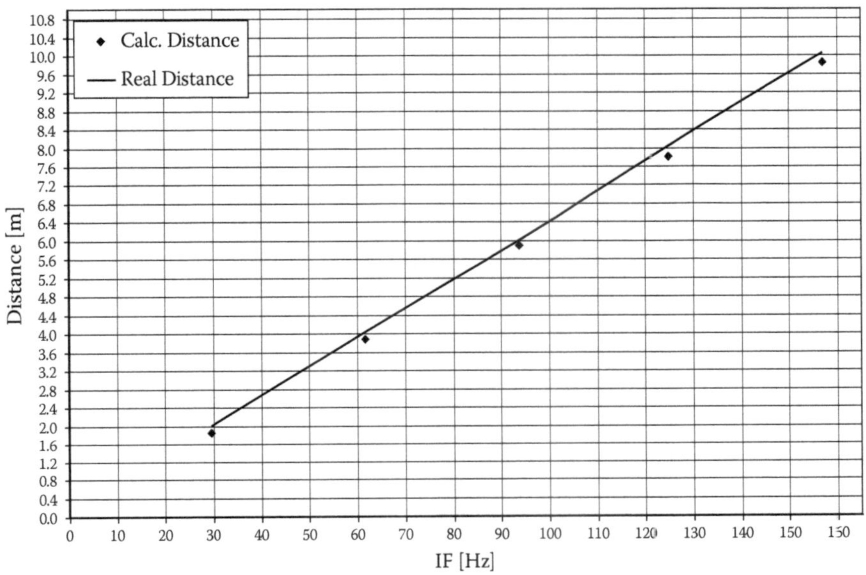

FIGURE 4.12
Comparison of distance as a function of the measured IF. (From Kaplevich, B. et al., in *IEEE MTT-S International Microwave Workshop Series on Millimeter Wave Integration Technologies*, Sitges, Spain, 2011, pp. 664–667. © 2011 by the Institute of Electrical and Electronics Engineers [IEEE].)

and covered by fabric. Figure 4.14 shows the configuration of plastic sticks and the measured results. We can clearly see the response of the IF consisting of the three peaks, each corresponding to a single stick.

4.3.4 Target Identification

The real environment is a source of various clutter which is responsible for multiple responses in the IF channel so that target identification must be done carefully in order to improve detection probability. Also, parasitic effects such as coupling between T_x and R_x antennae and leakage from LO due to insufficient isolation may cause additional low-frequency spectral components, screening the signal from a real target. Consequently, the IF spectra are strongly dependent on contributions of the above factors which may be different for indoor and outdoor scenarios. Figure 4.15 illustrates the IF spectra of a gun placed within a long corridor. The two IF responses (IF_1 = 65.795 Hz and IF_2 = 335 Hz) corresponding to 3 and 20 m are prominent. In order to identify each return, an additional absorbing sheet was placed behind the gun. The absorbing sheet suppresses parasitic reflection from the wall behind a target at a distance of about 20 m, as illustrated by Figure 4.16. In deployment, e.g. at a security checkpoint, such an arrangement must be carefully performed in order to minimize the false alarm rate caused by parasitic reflections.

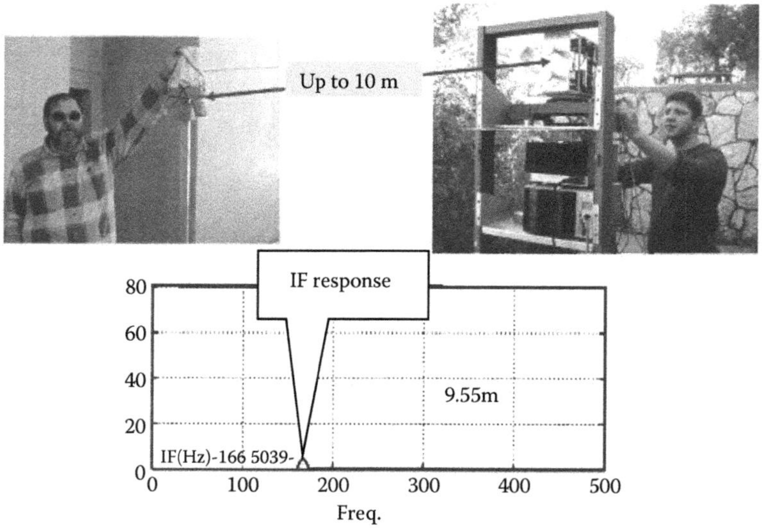

FIGURE 4.13
An outdoor experiment: detecting a gun hidden under fabric. (From Kapilevich, B. et al., in *IEEE MTT-S International Microwave Workshop Series on Millimeter Wave Integration Technologies*, Sitges, Spain, 2011, pp. 664–667. © 2011 by the Institute of Electrical and Electronics Engineers [IEEE].)

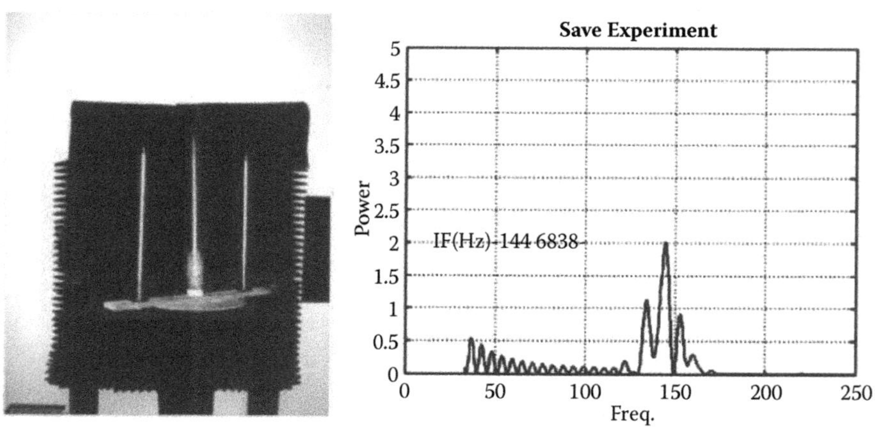

FIGURE 4.14
Experiment with a 94 GHz sensor detecting plastic sticks covered by fabric at a distance of about 3 m. (From Kapilevich, B. et al., in *IEEE MTT-S International Microwave Workshop Series on Millimeter Wave Integration Technologies*, Sitges, Spain, 2011, pp. 664–667. © 2011 by the Institute of Electrical and Electronics Engineers [IEEE].)

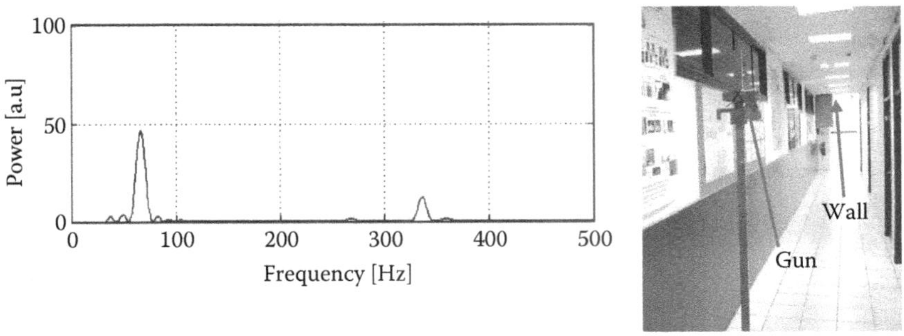

FIGURE 4.15
The measured IF and the optical picture of the gun and the wall behind it.

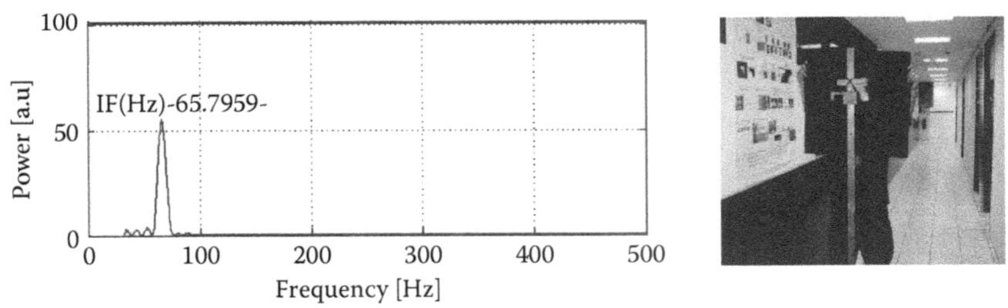

FIGURE 4.16
The measured IF and the optical picture of the gun in front of the absorbing sheet.

The situation becomes more complicated when a sensor operates in outdoor environments. Since clutter is often non-stationary, for example, backscattering from leaves on trees, more sophisticated processing algorithms should be used. One method to do this removes distortions, caused by hardware, from the measured data. Figure 4.17 illustrates a typical structure of IF voltage as a function sampling number synchronized with sawtooth modulation when the R_x antenna is blocked by an absorber. A fragment of the recorded signal within the selected chirp limits has been smoothed using cubic spline fitting as shown in Figure 4.18, then subtracted from the total IF signal corresponding to the real environment in the post-processing procedure.

As an example, Figure 4.19 demonstrates scenery used in outdoor experiments that consist of two vertical metal rods of diameter 8 cm and a horizontal section of aluminium power electric line of diameter 2.5 cm placed approximately 8 m from the sensor. The cars are approximately 14 and 13 m from the sensor, respectively. Bushes located approximately 20 m from the sensor form a typical natural background. The sensor is mounted on the Az-El positioner, and a single horizontal scan was performed over a field of view of about 40°. The signal in the IF channel was recorded as shown in Figure 4.20a. After applying the FFT procedure to a particular chirp, the IF spectral pattern has been restored as shown in Figure 4.20b, where the

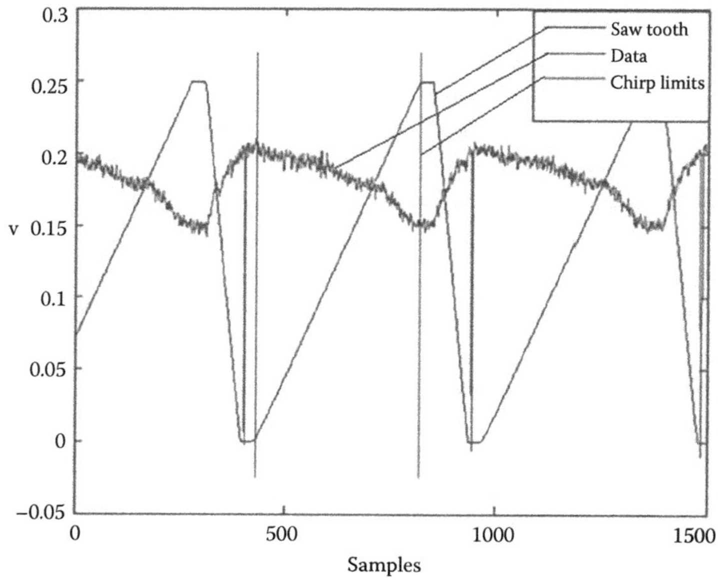

FIGURE 4.17
The typical structure of the IF voltage with a sample number when the R_x antenna is blocked by an absorber. (From Kapilevich, B. et al., in *IEEE MTT-S International Microwave Workshop Series on Millimeter Wave Integration Technologies*, Sitges, Spain, 2011, pp. 664–667. © 2011 by the Institute of Electrical and Electronics Engineers [IEEE].)

vertical axis corresponds to spectral density power in relative units (r.u.) and the horizontal axis is a distance from the sensor. Finally, the refined spectral pattern was obtained by subtracting the 'instrumental' part as shown in Figure 4.20c. In order to identify the targets we have drawn the plot shown in Figure 4.21 with the measured distances as a function of each pixel's number in the horizontal direction. We can clearly identify the left and right metal posts, cars and background clutter.

In order to detect the horizontal power electric line (cable of diameter 2.5 cm) shown in Figure 4.19, the additional vertical scan was performed within a field of view 10°. The recorded voltage of a signal in the IF channel corresponding to this case is depicted in Figure 4.22a. Repeating the aforementioned procedure, the result of target identification is presented in Figure 4.22b. Both the cable (at a distance of about 8 m) and bushes (at a distance of about 20 m) are clearly resolved.

Typically, real environment demonstrates a wide range of reflectance signatures, for example, bushes in the above experiments. These are weaker in comparison with the reflections from cars. Further, these weak reflections are unstable since positions of bush's trunks and leaves are in a permanent state of movement depending on wind velocity, creating an additional clutter effect. Another source of instability is the intermodulation products in

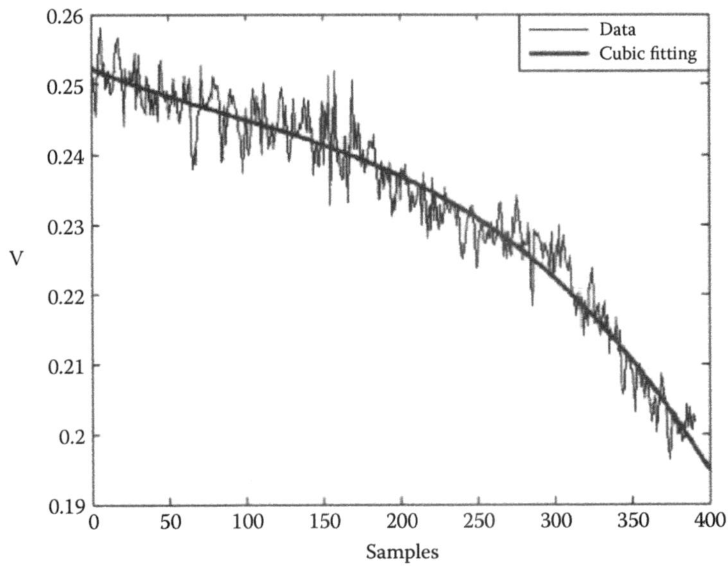

FIGURE 4.18
A part of the recorded signal and cubic fitting of the signal with blocked the R_x channel. (From Kapilevich, B. et al., in *IEEE MTT-S International Microwave Workshop Series on Millimeter Wave Integration Technologies*, Sitges, Spain, 2011, pp. 664–667. © 2011 by the Institute of Electrical and Electronics Engineers [IEEE].)

FIGURE 4.19
Scenery used in outdoor experiments. (From Kapilevich, B. et al., in *IEEE MTT-S International Microwave Workshop Series on Millimeter Wave Integration Technologies*, Sitges, Spain, 2011, pp. 664–667. © 2011 by the Institute of Electrical and Electronics Engineers [IEEE].)

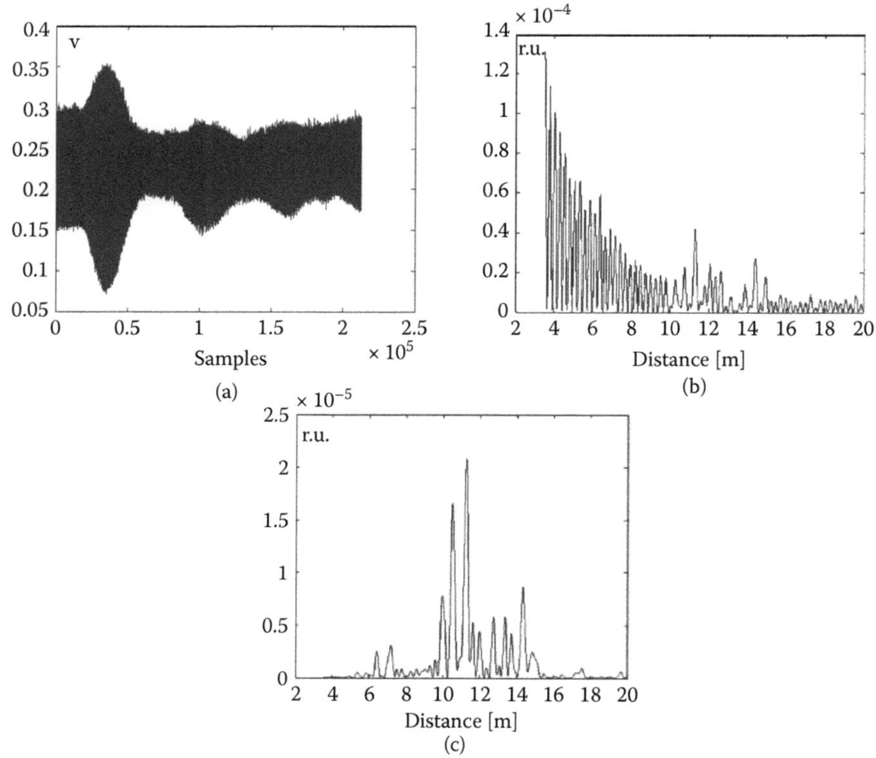

FIGURE 4.20
(a) The recorded voltage in the IF channel corresponding to the single horizontal scan, (b) the corresponding IF spectral pattern of the recorded signal, and (c) the refined IF spectral pattern. (From Kapilevich, B. et al., in *IEEE MTT-S International Microwave Workshop Series on Millimeter Wave Integration Technologies*, Sitges, Spain, 2011, pp. 664–667. © 2011 by the Institute of Electrical and Electronics Engineers [IEEE].)

the IF spectra. The above factors lead to uncertainty in distance estimation when the intensity of the received signal is close to the noise floor.

In order to detect a human body at distances of 50–100 m and more in adverse weather conditions (fog, smoke, dust, etc.), the sensor shown in Figure 4.9 has been modified by replacing horn-lens antennae by Gaussian beam antennae with the better performance (gain 35 dBi and 3° beamwidth at the 3 dB level) model QuinStar QLA-WOOYO3C. Also, the synthesizer was replaced by a voltage-controlled oscillator (VCO) modulated by a triangular wave signal, resulting in a chirp with linear frequency modulation at a frequency of 15.6–15.8 GHz. To create the chirp signal, an electronic circuit producing a 33 Hz triangular wave within the voltage range of 8–9.2 V was assembled. Figure 4.23 shows the general view of the modified sensor.

A typical outdoor scenario used in experiments is shown in Figure 4.24a. An integration time of 100 ms was employed, and during this time, three periods

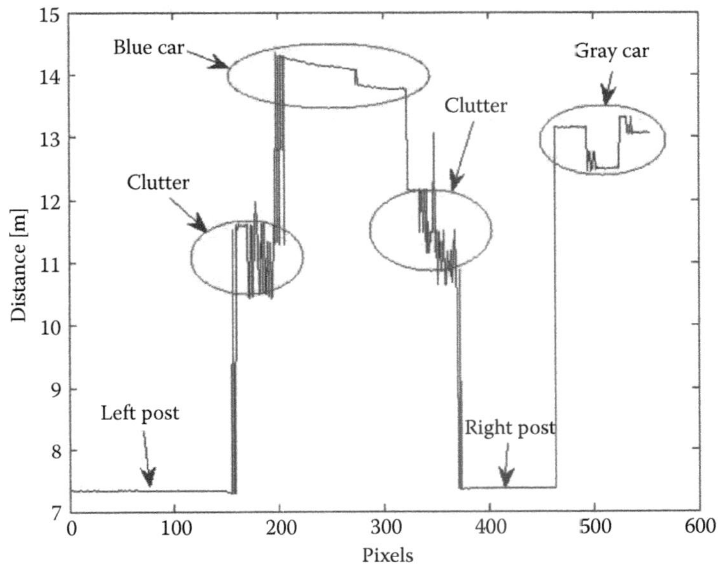

FIGURE 4.21
The measured distance as a function of a pixel's number for the horizontal scan. (From Kapilevich, B. et al., in *IEEE MTT-S International Microwave Workshop Series on Millimeter Wave Integration Technologies*, Sitges, Spain, 2011, pp. 664–667. © 2011 by the Institute of Electrical and Electronics Engineers [IEEE].)

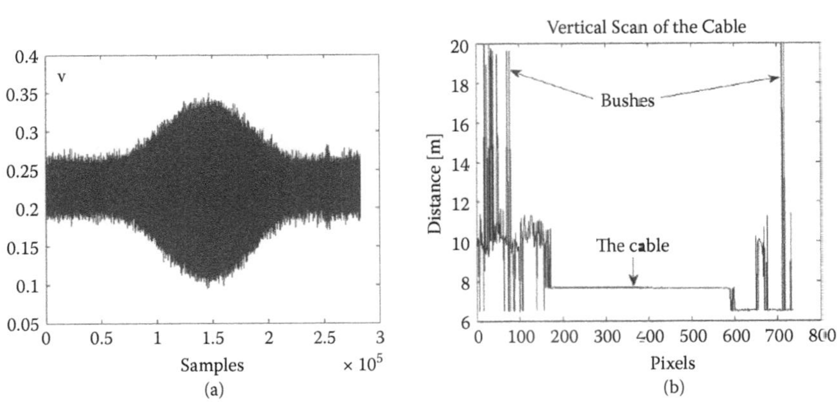

FIGURE 4.22
(a) The recorded voltage of a signal in the IF channel corresponding to a vertical scan and (b) the measured distance as a function of the pixel's number for the vertical scan. (From Kapilevich, B. et al., in *IEEE MTT-S International Microwave Workshop Series on Millimeter Wave Integration Technologies*, Sitges, Spain, 2011, pp. 664–667. © 2011 by the Institute of Electrical and Electronics Engineers [IEEE].)

FIGURE 4.23 (See color insert.)
The modified sensor. (From Etinger, A. et al., *IEEE Sensors Journal*, 14, 4, 1232–1237, 2014. © 2014 by the Institute of Electrical and Electronics Engineers [IEEE].)

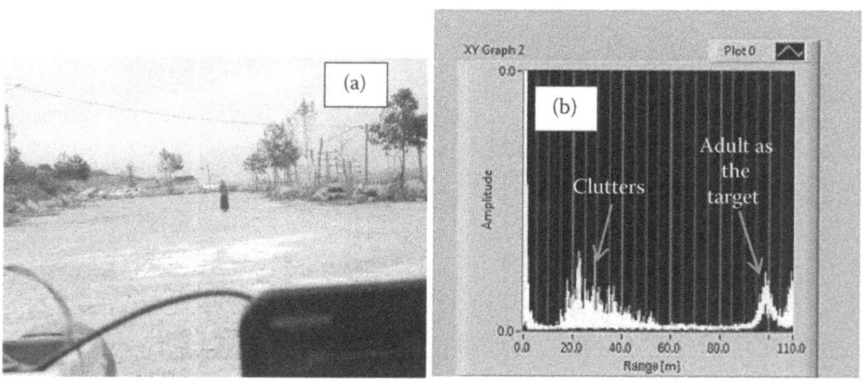

FIGURE 4.24
(a) The typical scenario in outdoor experiments and (b) the screenshot of human detection at a distance about 90 m using the W-band sensor. (From Etinger, A. et al., *IEEE Sensors Journal*, 14, 4, 1232–1237, 2014. © 2014 by the Institute of Electrical and Electronics Engineers [IEEE].)

of the triangular modulating wave were integrated, allowing detection of an adult human at a distance of about 90 m. The screenshot of the LabView interface is shown in Figure 4.24b and illustrates the situation described.

4.4 Submillimetre-Wave Sensor at 330 GHz

The choice of operating frequency is a critical issue for mm-wave sensors. Indeed, mm-wave can penetrate to a reasonable depth through various materials (far better than IR radiation) and result in mm-scale resolution in the imaging process (better than microwave frequencies). However, mm-wave components are more expensive than their microwave counterparts. In addition, microwave receivers are more sensitive and demonstrate better noise performance but suffer from lower resolution. Submillimetre-wave imagers have better resolution than mm-wave versions, but they suffer from a lower sensitivity still. There are limited data about submillimetre-wave sensors at present, and a systematic study of their application for remote detection of hidden objects will help designers in finding the key trade-offs between resolution and sensitivity.

4.4.1 Schematic of the Sensor

One possible configuration of a submillimetre-wave sensor is shown in Figure 4.25. The HP-8350B sweep oscillator is used as the source of an X-band FMCW signal. Several multiplying stages are used in the R_x and T_x channels. In order to drive the sub-harmonic mixer in the R_x channel, the multiplication factor is chosen as 16, while in the T_x channel this factor is 32, providing FMCW radiation at near 330 GHz with an output power of about 10 mW. Power amplifiers (PAs) are employed in both channels in order to optimize multiplier performance. The experimental setup is assembled from off-the-shelf, solid-state components manufactured by Virginia Diodes Ltd., Quinstar, Spacek, and Millitech. Table 4.2 summarizes the list of basic components employed in the setup shown in Figure 4.26.

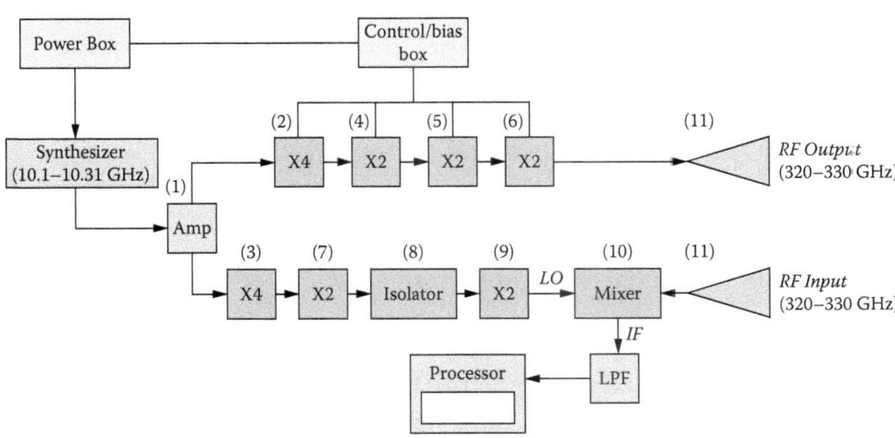

FIGURE 4.25
The functional diagram of the 330 GHz submillimetre-wave sensor. (From Kapilevich, B. et al., in *IEEE Conference on Microwaves, Communications, Antennas, and Electronics Systems (COMCAS)*, 1–4, 2011. © 2011 by the Institute of Electrical and Electronics Engineers [IEEE].)

TABLE 4.2

330 GHz Millimetre-Wave Sensor: Basic RF Components List

Index	Component
1	Amplifier QuinStar QPJ-06183040-AO
2	Amplifier and 4× multiplier Spacek A425-4XWB-25
3	Amplifier and 4× multiplier Spacek A1040-4XWB-23
4	2× multiplier VDI D80v4-S034b
5	2× multiplier VDI D162-S021c
6	2× multiplier VDI D320r4-S1-63b
7	2× multiplier VDI WR10xc-S14d
8	Isolator Millitech FBI-10-RSEB0
9	2× multiplier VDI WR5.1x2-S1-65b
10	2nd harmonic mixer VDI WR2.8SHM-S1-29
11	22 dBi horn antennae

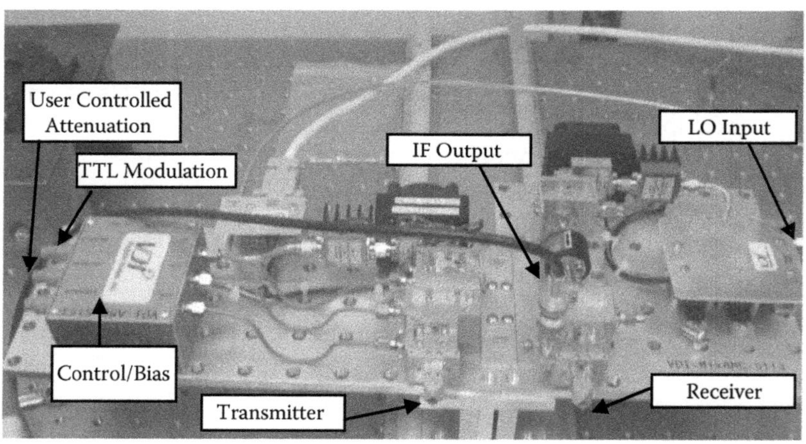

FIGURE 4.26
General view of the 330 GHz submillimetre-wave sensor with horn antennae.

4.4.2 System Characteristics and Performance

4.4.2.1 Components Parameters at 330 GHZ

Horn antennae gain: $G_r = G_t = 22$ dBi (158.5)

Mixer type WR2.8SHM: According to the component datasheet [5], the sub-harmonic mixer has a conversion loss of $CL_{mix} = 5.5$ dB (3.55).

The insertion loss of LPF is $F_{LPF} = 0.5$ dB (1.122) and gain $G_{LPF} = -0.5$ dB (0.891).

The video amp has $F_{video} = 2$ dB (1.595) and gain $G_{video} = 20$ dB (10^2).

4.4.2.2 Performance Estimation

The received power is derived from the radar equation. Assuming RCS = 10 cm² from a flat target illuminated by transmitter with output power P_t = 10 mW at a distance of 10 m, the received power is

$$P_{r,\min} = \frac{P_t \cdot G_r \cdot G_t \cdot \lambda^2 \cdot RCS}{(4\pi)^3 \cdot d^4} = \frac{10^{-2} \cdot 158.49^2 \cdot (0.91 \cdot 10^{-3})^2 \cdot 10 \cdot 10^{-4}}{(4\pi)^3 \cdot 10^4} = 1.05 \cdot 10^{-12} W$$

Then, the total noise figure can be estimated, assuming that environment temperature is 290K:

$$F_{TOT} = CL_{mix} + \frac{F_{LPF} - 1}{CL_{mix}} + \frac{F_{amp} - 1}{CL_{mix} \cdot G_{LPF}} + ... = \left(3.55 + \frac{1.122 - 1}{3.55} + \frac{100 - 1}{3.55 \cdot 1.122} \right.$$

$$= 26.4 \cong 14.2 \text{ dB}$$

$$T_{eqv} = T_0(F_{TOT} - 1) = 290(26.4 - 1) = 7366K$$

The expected signal-to-noise ratio at the receiver's output is found to be

$$\text{SNR} = 10\log\left(\frac{P_{r,\min}}{k_B T_{eqv} BW}\right) = 10\log\left(\frac{1.05 \cdot 10^4}{1.38 \cdot 10^{-23} \cdot 7366 \cdot 2.5 \cdot 10^4}\right)$$

$$= 10\log(41.3) \cong 16.1 \text{ dB}$$

The noise floor is found to be

$$NF = k_B T_{eqv} BW = 1.38 \cdot 10^{-23} \cdot 7366 \cdot 2.5 \cdot 10^4 = -125.6 \text{ dBm}$$

As a result,

$$MDS = -125.6 + 16.1 = 109.5 \text{ dBm}$$

Table 4.3 summarizes the basic operational parameters of the FMCW 330 GHz sensor.

4.4.3 Remote Detection Experiments

When performing distance measurements, a sweep time of 0.1 s yields the most consistent IF results, and therefore was chosen as an operational parameter. The gun covered by fabric at 1.41 m away from the sensor (Figure 4.27) yields an IF of 594.7 Hz and a calculated distance of 1.39 m, roughly translating to a 2 cm distance deviation.

The sampling rate, for processing of the IF output, is a critical parameter. It is illustrated by a comparison of the resulting distance resolution for the received and sampled signals using 10 and 100 KHz sampling rates. In

TABLE 4.3

Basic Operational Parameters of the FMCW 330 GHz Image Sensor

T_x Channel	
Transmitted power	10 mW
Frequency range	323.2–329.6 GHz
Frequency span	$\Delta f = 6.4$ GHz
Sweep time	$T_{sweep} = 0.1–0.2$ s
Transmitting antenna gain	22 dBi
R_x Channel	
Receiving antenna gain	22 dBi
Mixer conversion loss	6–7 dB
A/D sampling rate	10–100 KHz
Video amplifier gain	20–40 dB (adjustable)

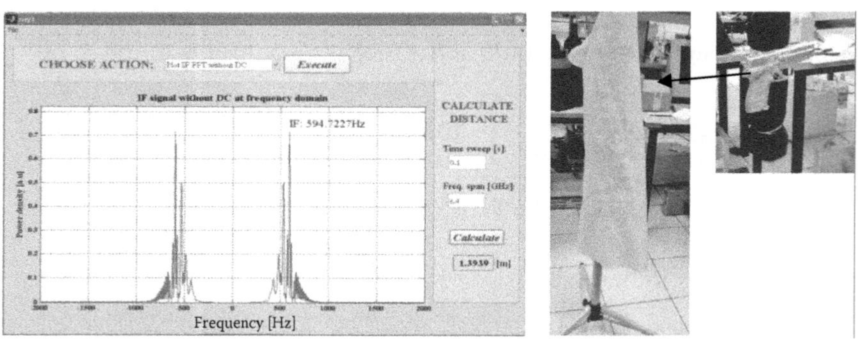

FIGURE 4.27
The IF signal and the concealed metal gun.

order to estimate the actual role of this parameter, a distance measurement to a metal plate placed perpendicularly to the incident wave at a distance 1.67 m from the sensor was taken 15 times with a sweep time of 0.1 s. The results are as follows: using the 10 KHz sampling rate DAQ yields different results with a standard deviation of 1.05 cm and an average value of 1.6525 m (Figure 4.28a), while sampling the same signals using the 100 KHz DAQ produced an exact calculated distance of 1.6522 m (Figure 4.28b).

4.4.4 Improving Sensitivity and Resolution

In order to improve sensitivity and resolution, each horn antenna was integrated with a dielectric focusing lens. The dielectric constant of the lens and its position with respect to the phase centre of the horn should be chosen very accurately for optimum performance of the horn-lens antenna system.

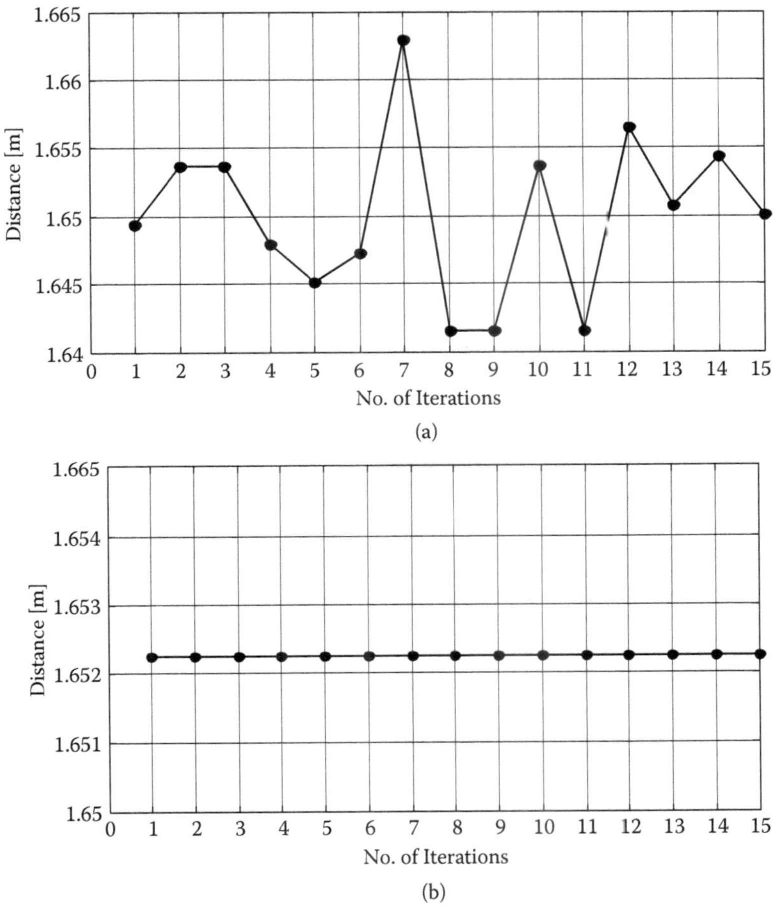

FIGURE 4.28
Calculated distance using the (a) 10 KHz A/D and (b) 100 KHz A/D.

The general geometry of a typical horn-lens configuration is shown in Figure 4.29. Assuming that the phase centre of a horn antenna is placed in the focal point of a dielectric lens, the rule of thumb for choosing the basic dimensions can be formulated as follows [6]:

$$F = \frac{dr\pi}{4\lambda} \tag{4.13}$$

where F is the lens focal length, d is the beam diameter at the focus, and r is the $1/e^2$ diameter of the collimated beam. In order to demonstrate the critical role of the basic dimensions and dielectric constant of lens material, a 3D EM simulation of the directivity, main lobe beamwidth and side lobe level was performed using the solver CST Microwave Studio. The simulations have

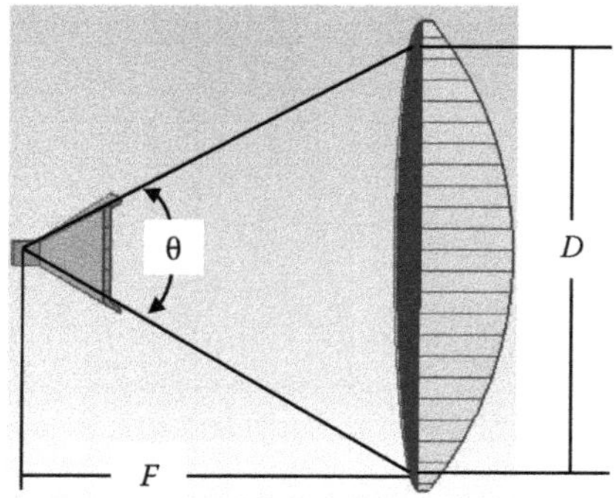

FIGURE 4.29
The model of horn-lens system used in CST simulations.

been made at a frequency of 40 GHz assuming that $\theta = 30°$, $F = D = 60$ mm, and relative dielectric constant of the lens material of 3. The effect of variations of the angle θ is shown in Figure 4.30. With increasing θ, directivity D is monotonically increased, while the 3 dB beamwidth (BW) is monotonically decreased. The longitudinal shift of the horn's phase centre with respect to the focal point along the axis of focusing lens, denoted as *delta* (in wavelengths), may drastically degrade overall system performance. Figure 4.31 depicts the role of this parameter, which must be chosen very accurately to ensure good performance of the horn-lens system.

The effect of variations of the relative dielectric constant ε_r is illustrated by Figure 4.32. The maximum directivity, minimum beamwidth and side lobe level correspond to $\varepsilon_r \sim 3$ for the configuration shown in Figure 4.29. Finally, Figure 4.33a shows comparisons of the 2D field radiation patterns with the focusing lens and without it. A comparison of 3D field radiation patterns is presented in Figure 4.33b. Adding a focusing lens is a simple way to increase the total directivity by about 10 dB and decrease 3 dB beamwidth about two to three times, resulting in improved sensitivity and spatial resolution of the submillimetre-wave sensor. Other related topics concerning designing a horn-lens focusing system can be found in [8]–[10].

The two modifications of the submillimetre-wave sensor have been assembled and tested. The first one (Figure 4.34a) employs the Teflon plano-convex focusing lens of diameter 50 mm with a focal length of 100 mm (model LAT-100, Thorlabs [7]). The two longitudinal slots allow adjusting the positions of lenses with respect to the apertures of the horn antennae in order to obtain the minimum size of beam spot. By proper adjusting of each lens position, the measured diameter of a beam spot (3 dB level) is about 25–30 mm at a distance of 1.6 m, as shown in Figure 4.34b.

In the second modification the horn antennae were replaced by horn-lens Gaussian beam antennae, as shown in Figure 4.35a. The general view of the sensor with these antennae is depicted in Figure 4.35b.

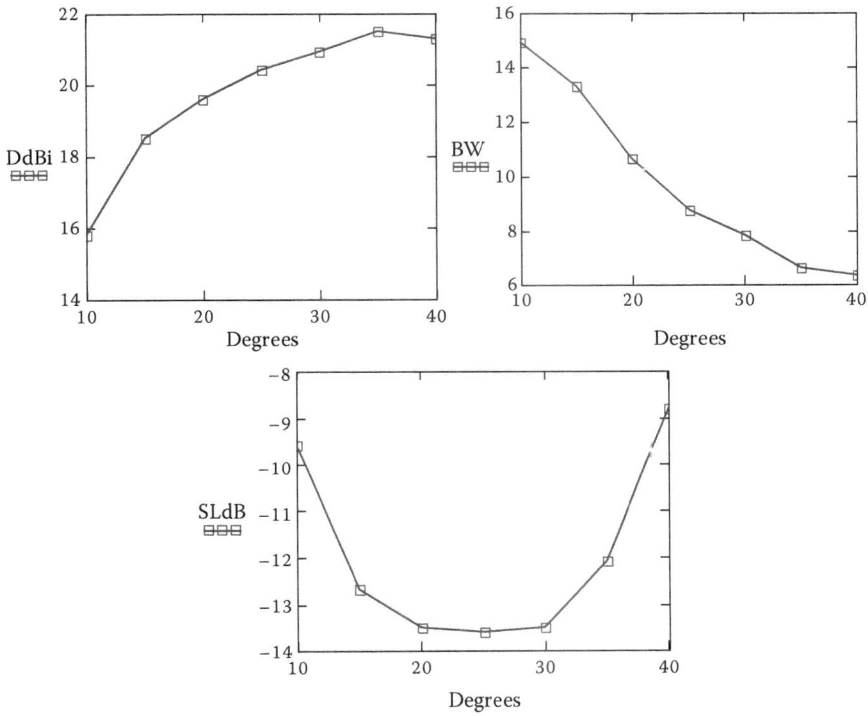

FIGURE 4.30
Variations of the directivity (D, dBi), 3 dB beamwidth (BW, degrees) and side lobe level (SL, dB) as a function of the flare angle θ.

Integration of horn antennae with lenses is widely used in radiating systems to collimate the radiation from a broader angle source like a horn into the narrower beam. Such horn-lens antennae have an advantage over reflector antennae because of the absence of aperture blockage, resulting in an increase in efficiency and a reduction of side lobe levels. For a Teflon lens, with a relative dielectric constant of 2.08 and $F/D = 0.93$, the required edge taper for good performance is –5 dB [11]. The conical horn had an aperture diameter of $D = 7$ mm that produces almost equal E- and H-plane beamwidths. It was designed to illuminate the lens with an edge taper of –15 dB, corresponding to the beamwidth constant of about 72, using the nomenclature described in [12]. An estimation of the directivity at 330 GHz is given by

$$G = \eta\left(\frac{\pi D}{\lambda}\right)^2 \cong 42 \text{ dBi} \tag{4.14}$$

where $\eta = 0.55$, the aperture efficiency, and $D = 52$ mm, the aperture diameter.

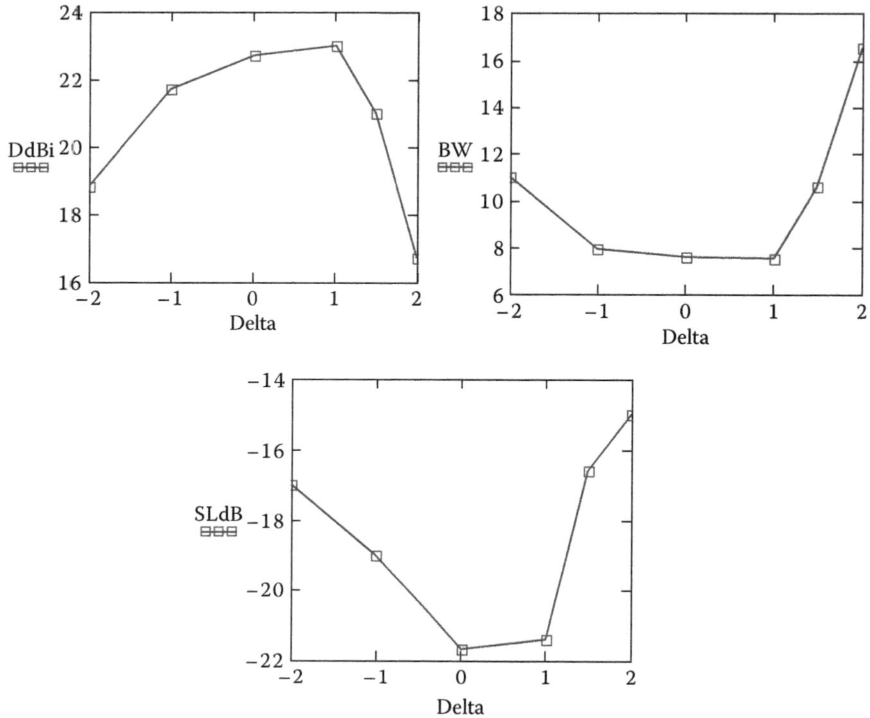

FIGURE 4.31
Variations of the directivity (D, dBi), 3 dB beamwidth (BW, degrees) and side lobe level (SL, dB) as a function of the parameter *delta* in wavelengths.

The beamwidth of the antenna is approximated by

$$\theta_{beam} = 4\lambda/(\pi D) \cong 1.3°.$$

The dielectric-air boundaries are the source of multiple reflections leading to increased side lobe levels. The average loss due to reflections can be approximated as follows:

$$\alpha_r = 8.69\left(\frac{\sqrt{\varepsilon_r} - 1}{\sqrt{\varepsilon_r} + 1}\right)^2 \tag{4.15}$$

Assuming that ε_r = 2.08, the average attenuation due to reflection from the single surface is about 0.285 dB. If both surfaces interfere independently in-phase, the total reflection is duplicated, leading to four times increasing power attenuation due to multiple reflections of up to 1.1 dB in our case. This treatment is valid if partial reflections are small enough. Another factor

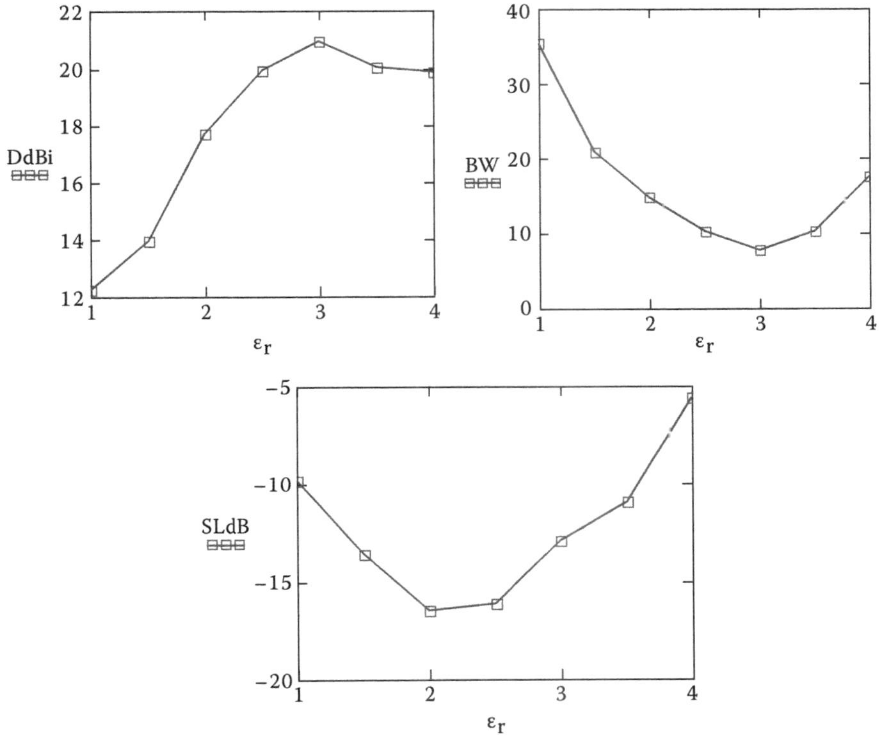

FIGURE 4.32
Variations of the directivity (D, dBi), 3 dB beamwidth (BW, degrees), and side lobe level (SL, dB) as a function of the relative dielectric constant, ε_r.

responsible for losses is attenuation in dielectric lens. The attenuation in dB per unit wavelength through the lens is given by

$$\alpha_d = 27.3\sqrt{\varepsilon_r}\tan(\delta) \tag{4.16}$$

For Teflon the loss tangent is $\tan(\delta) = 5 \times 10^{-4}$, and for a maximum lens thickness of 15 mm, the total attenuation due to dielectric loss is about 0.3 dB, resulting in a total attenuation through lens of about 1.4 dB. The gain is modified by this loss to produce the final antenna gain of 40.3 dBi. The horn-lens antennae have been fabricated by ELVA [13]. The measured radiation pattern of this horn-lens antenna in E- and H-planes is shown in Figure 4.36. The measured 3 dB beamwidth and gain are in good agreement with the calculated parameters above.

The high-gain antennae allow detection of objects with a small RCS up to a range of several tens of metres [15]. At these distances the expected IF signals are within the kHz or tens of kHz frequency bands. Figure 4.37a shows the corner reflector as a test target, having 35 cm length of its greater dimension. The reflector was located at a distance of 24.5 m from the sensor. The

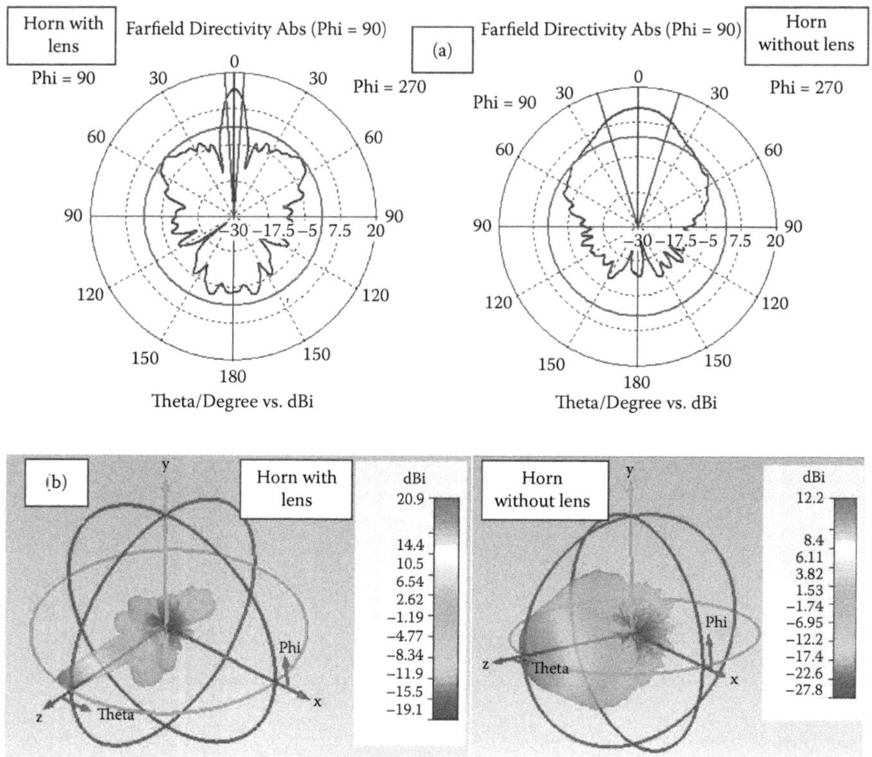

FIGURE 4.33
Calculated 2D and 3D field radiation patterns with the focusing lens and without it.

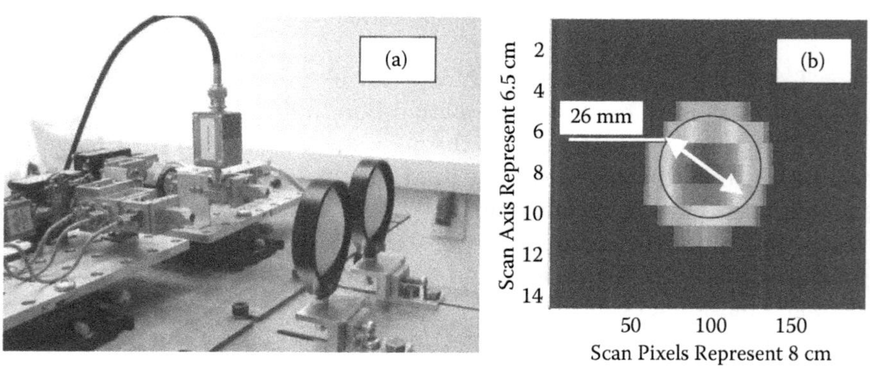

FIGURE 4.34
(a) A view of the modified 330 GHz sensor with the focusing lens and (b) the measured diameter of the beam spot.

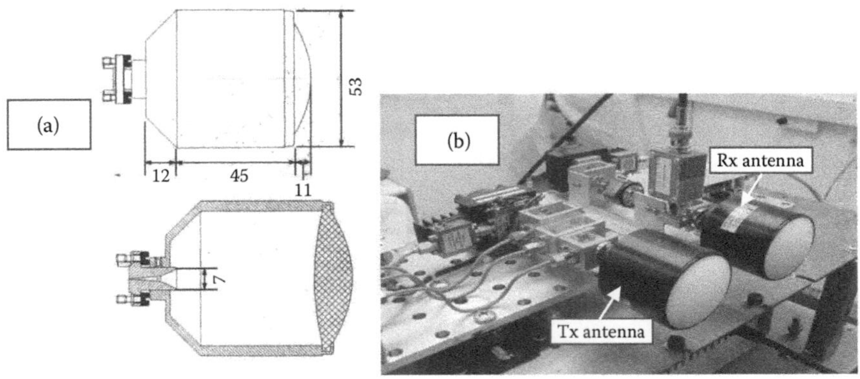

FIGURE 4.35 (See color insert.)
(a) The horn-lens Gaussian beam antenna and (b) general view of 330 GHz sensor with these antennae. (From Kapilevich, B. et al., in *IEEE Conference on Microwaves, Communications, Antennas and Electronics Systems (COMCAS)*, 1–4, 2011. © 2011 by the Institute of Electrical and Electronics Engineers [IEEE].)

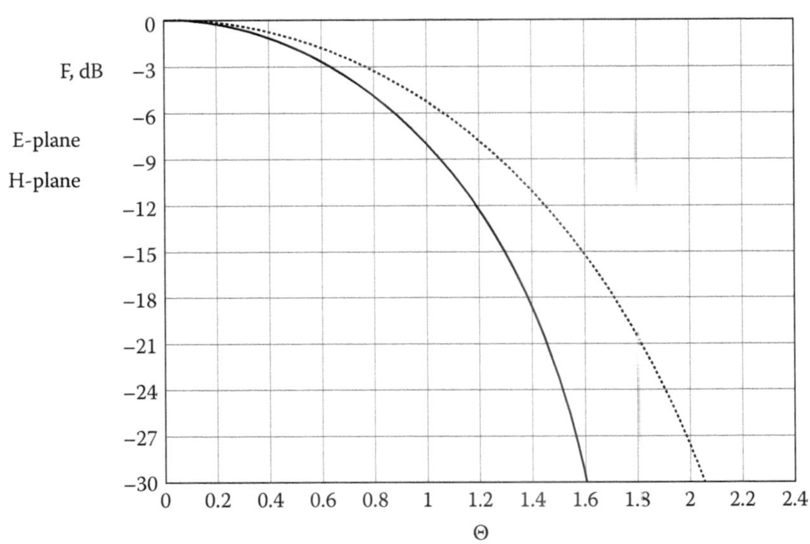

FIGURE 4.36
The measured pattern of a horn-lens antenna as a function of the elevation angle in degrees: solid line, E-plane; dotted line, H-plane.

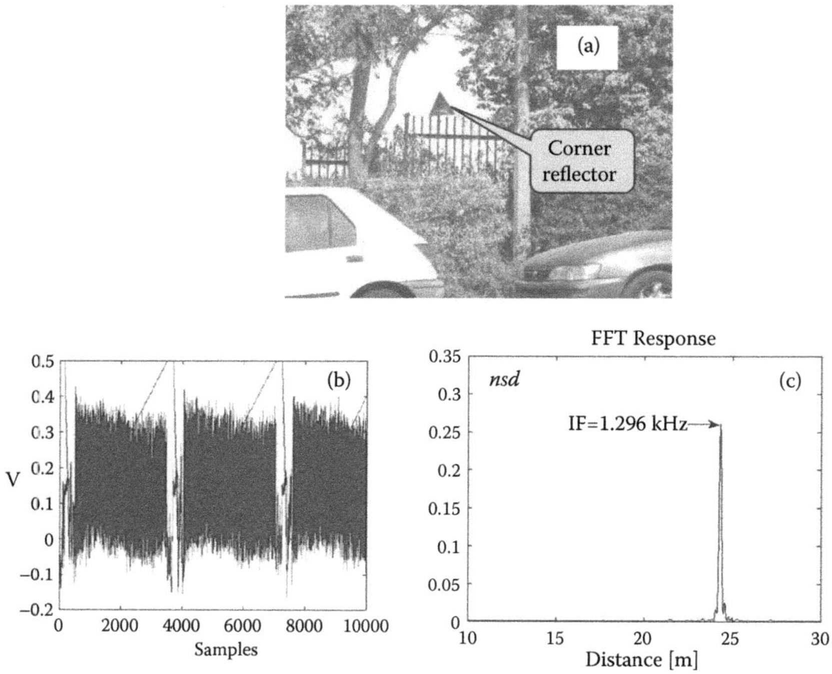

FIGURE 4.37
Tested target: (a) a corner reflector at distance 24.5m; (b) the IF signal in the time domain; (c) the resultant spectrum in the frequency domain after FFT and scaled to the distance from target, IF = 1.296 KHz. (From Kapilevich, B. et al., in *IEEE Conference on Microwaves, Communications, Antennas and Electronics Systems (COMCAS)*, 1–4, 2011. © 2011 by the Institute of Electrical and Electronics Engineers [IEEE].)

experiment has been carried out with $T_{sweep} = 0.2$ s and $\Delta f = 50$ MHz × 32 = 1.6 GHz (compression ratio $\Delta f/T_{sweep} = 8$ GHz/s). The signal recorded at the IF output is shown in Figure 4.37b. The normalized spectrum density (nsd) of this signal after applying the FFT procedure and scaled to the distance from the target is given in Figure 4.37c. Calculation of the measured distance has yielded $d = 24.3$ m, which is close to the actual distance and does not exceed the longitudinal measurement error predicted.

In order to estimate the spatial resolution of the designed sensor, a metal disk of 26 mm diameter has been installed on an absorbing background (Figure 4.38a). Assuming a far-field approximation, the beam diameter at a distance d from the transmitting lens is estimated as $W \approx d\ \theta_{beam} \approx 3.6$ cm at a distance of 1.6 m, where the signal-to-noise ratio is about 10 dB. The measured spot size on the –3 dB level is about 2 cm or less. So, the above estimation corresponds to the upper limit of spatial resolution. The reconstructed image of this target is shown in Figure 4.38b.

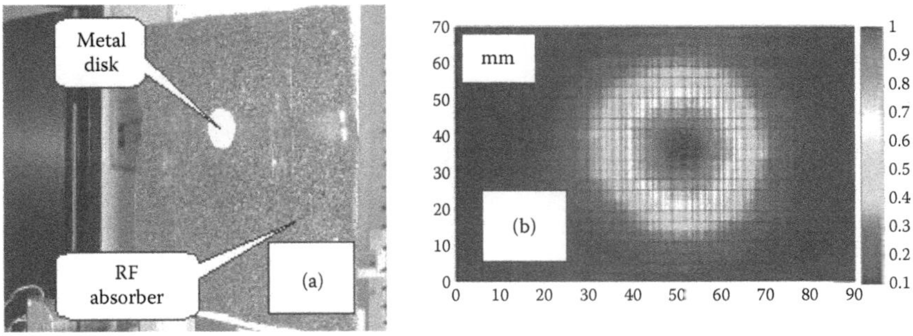

FIGURE 4.38
Optical (a) and submillimetre-wave (b) images of 26 mm diameter metal disk at 1.6 m distance.

The range resolution of the sensor has been estimated using a metal plate installed in front of the system and moved in the longitudinal direction. Signals in the IF port have been recorded for different distances and then transformed to the frequency domain using FFT. Finally, the increment of the distance to metal plate Δd has been determined using Equation 4.12. Theoretically, the increment in distance to the target and its measured value should be identical, leading to a straight line inclined by 45°, as shown in Figure 4.39. However, non-linearity of the FMCW synthesizer causes deviation from this line. In our case the measured maximum deviation is 1 cm, while the standard deviation from linear law is 3.6 mm [14].

The sensor was also used in experiments to detect high-voltage power lines, which create additional threats for helicopters and low-flying aircraft due to their low visibility. Experiments targeted on detecting such

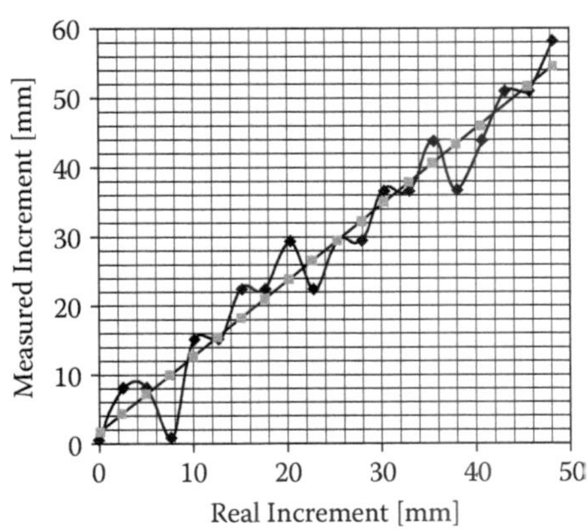

FIGURE 4.39
Comparison of the real and measured increments of distance to the target. (From Kapilevich, B. et al., in *IEEE Conference on Microwaves, Communications, Antennas, and Electronics Systems (COMCAS)*, 1–4, 2011. © 2011 by the Institute of Electrical and Electronics Engineers [IEEE].)

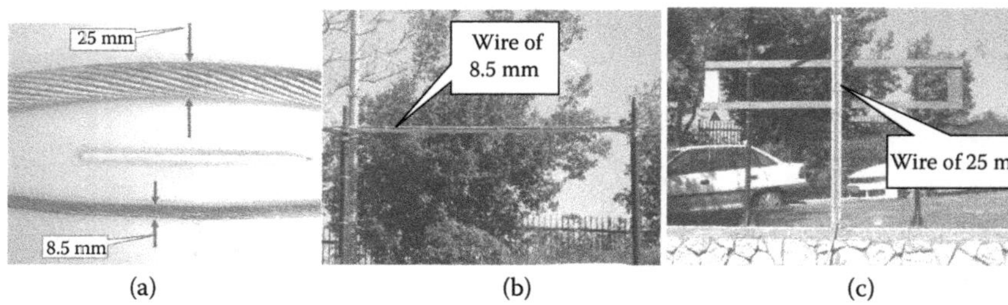

FIGURE 4.40 (See color insert.)
(a) The copper line used in a 35 KV power network (diameter is 8.5 mm) and the aluminium line used in a 160 KV power network (diameter is 25 mm). (b) The section of copper wire mounted on a special stand in the horizontal position. (c) The section of aluminium wire mounted on a special stand in the vertical position.

low-visible objects have been performed with the two types of power lines shown in Figure 4.40a: the cooper line used in the 35 KV power network, which has a diameter of 8.5 mm, and the aluminium line used in the 160 KV power network, which has a diameter of 25 mm. Sections of electric power lines of lengths of about 3 m were mounted on special stands in horizontal and vertical positions as depicted in Figure 4.40b,c. As a result, both horizontal and vertical polarization states of waves illuminating the targets can be compared in order to estimate the role of polarization effects [34].

The measured distance to the wires in all experiments is about 8 m, with accuracy less than 5 cm for both polarizations (Figure 4.41). However, the intensity of scattered signal strongly depends on polarization of the radiated wave, and a two-channel polarimetric receiver is desirable for improving overall sensor performance and increasing the range of detection as reported in [16]. Additional details concerning wire detection using mm-wave FMCW radar can be found in [26]–[28].

4.5 Conclusion

In this chapter, basic versions of FMCW sensors are presented for the detection of concealed objects. Fundamental equations governing the FMCW principle and important parameters needed for the optimal design of such sensors are explained. Examples of sensors operating at 94 and 330 GHz are given. Their basic parameters have been calculated and measured in real-world indoor and outdoor environments. The role of basic factors affecting the overall performance of the sensors designed for detecting concealed objects is discussed in detail. It should be noted that polarimetric versions of the sensors are able to estimate both co- and cross-polarized components backscattered by a target. They deliver additional information needed for

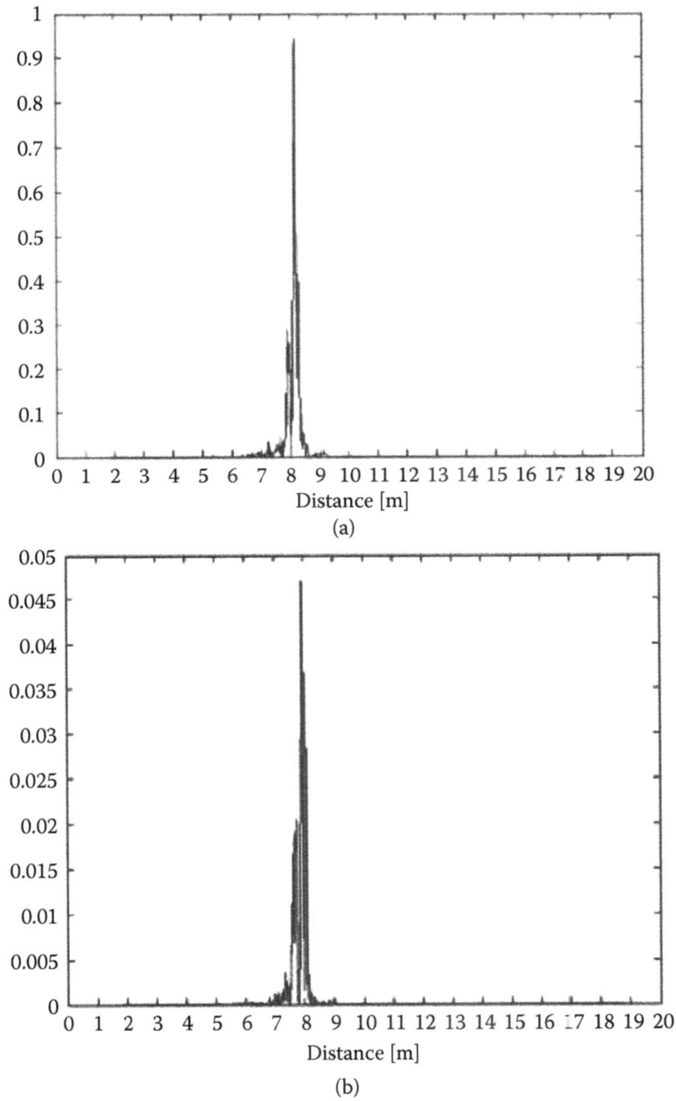

FIGURE 4.41
(a) The measured distances to wire in horizontal polarization with the diameter of copper wire 8.5 mm and (b) wire in vertical polarization with the diameter of aluminium wire 25 mm. The vertical axis corresponds to intensity of IF signal in relative units. (From Kapilevich, B. et al., in *IEEE Conference on Microwaves, Communications, Antennas and Electronics Systems* (COMCAS), 1–4, 2011. © 2011 by the Institute of Electrical and Electronics Engineers [IEEE].)

target identification and assist in the detection of concealed weapons carried by a human as well as assisting to distinguish a person from, for example, a dog [20]. The examples of various versions of FMCW radar sensors are reported in [21]–[25]. The reader may find in these references useful material which might be used to implement sensors intended for remote detection of concealed objects. Recent achievements in manufacturing mm-wave MMIC components open excellent possibilities for miniaturizing the next generations of such sensors [33, 34].

References

1. I. V. Komarov and S. M. Smolsky, *Fundamentals of short-range FM radar*, Artech House, Norwood, MA, 2003.
2. E. A. Wolff and R. Kaul, *Microwave engineering and systems applications*, Wiley, New York, 1988.
3. Y. Pinhasi, *Physical foundations of electronic communications*, Amichay, 2006.
4. B. Kapilevich, Y. Pinhasi, M. Anisimov, B. Litvak, and D. Hardon, FMCW MM-wave non-imaging sensor for detecting hidden objects, *IEEE MTT-S International Microwave Workshop Series on Millimeter Wave Integration Technologies, 2011*, Spain, pp. 664–667.
5. Available at http://www.vadiodes.com.
6. Available at http://www.eksmaoptics.com/en/main/formulas/useful.
7. Available at http://www.thorlabs.de/thorProduct.cfm?partNumber=LAT100.
8. J. Thornton and K.-C. Huang, *Modern lens antennas for communication engineering*, Wiley, New York, 2013.
9. O. Xu, Collimation lens design using AI-GA technique for Gaussian radiators with arbitrary aperture field distribution, *Journal of Electromagnetic Waves and Applications*, 25, 743–754, 2011.
10. E. Ongareau, E. Marouby, and J. R. Levrel, Charts for a quick design of spot-focusing corrugated horn lens antennas, in *Antennas and Propagation Society International Symposium*, 1994, pp. 986–989.
11. P. Bhartia and I. Bahl, *Millimeter wave engineering and applications*, John Wiley & Sons, New York, 1984.
12. R. Johnson, *Designer notes for microwave antennas*, Artech House, Norwood, MA, 1991.
13. Available at www.elva-1.com.
14. Y. Pinhasi, B. Kapilevich, R. Arusi, M. Anisimov, D. Hardon, B. Litvak, and Y. Wool, 330GHz FMCW image sensor for homeland security applications, *Journal of Infrared, Millimeter, and Terahertz Waves*, 31(11), 1370–1381, 2010.
15. B. Kapilevich, Y. Pinhasi, M. Anisimov, D. Hardon, B. Litvak, and Y. Wool, Detecting power lines using 330GHz FMCW sensor, in *IEEE Conference on Microwaves, Communications, Antennas and Electronics Systems (COMCAS)*, 2011, pp. 1–4.

16. K. Sarabandi and M. Park, Millimeter-wave radar phenomenology of power lines and a polarimetric detection algorithm, *IEEE Transactions on Antennas and Propagation*, 47(12), 1807–1813, 1999.
17. P. D. L. Beaslev and A. G. Stove, PILOT—an example of advanced FMCW techniques, in *IEE Colloquium on High Time-Bandwidth Product Waveforms in Radar and Sonar*, 1991, pp. 10/1–10/5.
18. N. Carrie and C. Broun, *Principles and applications of millimeter-wave radar*, Artech House, Norwood, MA, 1987.
19. G. M. Brooker, D. Birch, and J. Solms, W-band airborne interrupted frequency modulated CW imaging radar, *IEEE Transactions on Aerospace and Electronic Systems*, 41(3), 955–971, 2005.
20. M. Vahidpour and K. Sarabandi, Millimeter wave RCS and doppler spectrum of walking human and dog, in *Proceedings of IEEE Antennas and Propagation Society International Symposium*, 2007, pp. 4004–4007.
21. G. Brooker, *Introduction to sensors for ranging and imaging*, SciTech Publishing Inc., Raleigh, North Carolina, 2009.
22. H. J. Hansen, R. Lindop, and D. Majstorovic, *Collision avoidance W-band FMCW radars in an altimeter application*, Report DSTO-TR-1939, Australian Government Department of Defense, 2006.
23. J. B. Mead, I. Popstefanija, P. Kollias, B. Albrecht, and R. Bluth, Compact airborne solid-state 95 GHz FMCW radar system, *in 31st International Conferenceon Radar Meteorology*, paper 4A.3.
24. W. Mayer, M. Meilchen, W. Grabherr, P. Nüchter, and R. Gühl, Eight-channel 77-GHz front-end module with high-performance synthesized signal generator for FM-CW sensor applications, *IEEE Transactions on Microwave Theory and Techniques*, 52(3), 993–1000, 2004.
25. A. B. del Campo, A. A. Lopez, B. P. D. Naranjo, J. G. Menoyo, D. R. Moran, C. C. Duarte, and J. L. J. Martın, Millimeter-wave radar demonstrator for high resolution imaging, in *European Radar Conference*, Amsterdam, 2004, pp. 65–68.
26. K. Sarabandi, Millimeter-wave radar phenomenology of power lines and a polarimetric detection algorithm, *IEEE Transactions on Antennas and Propagation*, 47(12), 1807–1813, 1999.
27. C. J. Cornelio and K. N. Crocker, Helicopter electronic system for detecting electric power lines during landing, U.S. Patent 5859597, 12 January 1999.
28. K. Sarabandi, A radar cross-section model for power lines at millimeter-wave frequencies, *IEEE Transactions on Antennas and Propagation*, 51(9), 2363–2360, 2003.
29. G. M. Brooker, Understanding millimetre wave FMCW radars, in *1st International Conference on Sensing Technology*, New Zealand, 2005, pp. 152–157.
30. C. Bactir, E. Sobaci, and A. Donmez, A guide to reduce the phase noise effect in FMCW Radars, in *IEEE Radar Conference*, 2012, pp. 0236–0239.
31. K. S. Kulpa, Novel method of decreasing influence of phase noise on FMCW radar, in *Proceedings of CIE International Radar Conference*, 2011, pp. 319–323.
32. P. Beasley, The influence of transmitter phase noise on FMCW radar performance, in *3rd European Radar Conference, EuRAD-2006*, pp. 331–334.
33. S. Y. Kim, O. Inac, C.-Y. Kim, D. Shin, and G. M. Rebeiz, A 76–84-GHz 16-element phased-array receiver with a chip-level built-in self-test system, *IEEE Transactions on MTT*, 61(8), 3083–3098, 2013.

34. F. Golcuk, T. Kanar, and G. M. Rebeiz, A 90–100-GHz 4 × 4 SiGe BiCMOS polarimetric transmit/receive phased array with simultaneous receive-beams capabilities, *IEEE Transactions on MTT*, 61(8), 3099–3114, 2013.
35. A. Etinger, N. Balal, B. Litvak, M. Einat, B. Kapilevich, and Y. Pinhasi, Non-imaging mm-wave FMCW sensor for pedestrian detection, *IEEE Sensors Journal*, 14(4), 1232–1237, 2014.

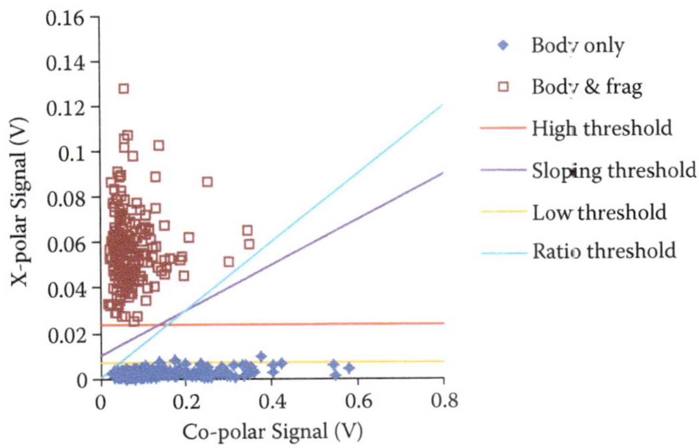

COLOR FIGURE 3.15
Measured co-polarized and cross-polarized signal amplitudes from 200 radar returns resulting from the human torso at a range of ~10 m. The person is either without objects or has a simulated PBIED. (From Harmer, S. W. et al., *IEEE Sensors Journal*, 99, 2013, doi: 10.1109/JSEN.2013.2273437.)

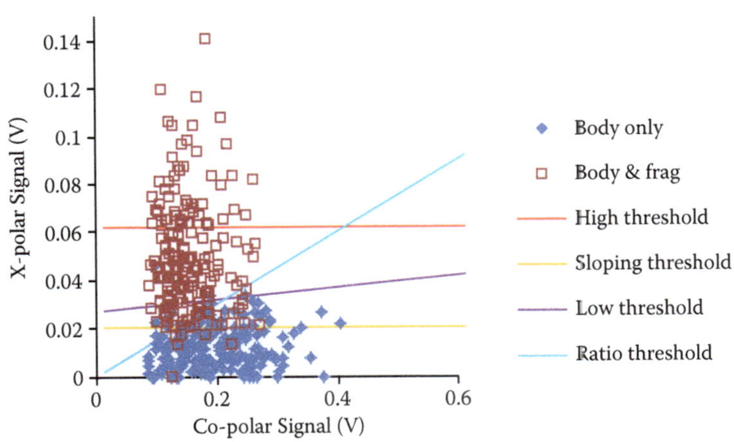

COLOR FIGURE 3.16
Measured co-polarized and cross-polarized signal amplitudes from 200 radar returns resulting from the human torso at a range of ~20 m. The person is either without objects or has a simulated PBIED. (From Harmer, S. W. et al., *IEEE Sensors Journal*, 99, 2013, doi: 10.1109/JSEN.2013.2273487.)

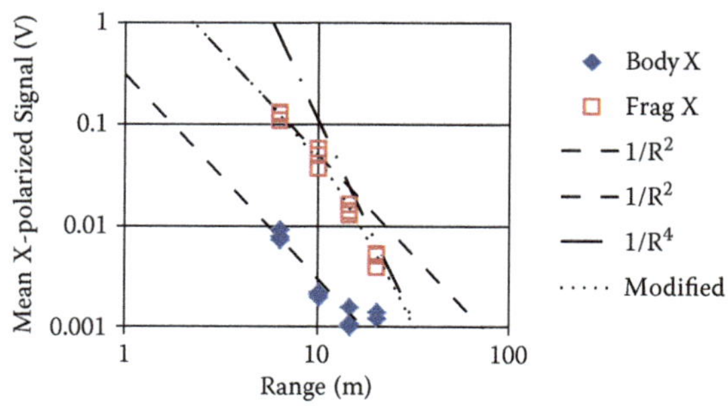

COLOR FIGURE 3.17

A log–log plot of the mean value of the measured cross-polarized signal amplitude (V) from the human body as a function of the range to the body. The body has no objects or has an attached PBIED. The inverse square range dependence is shown for comparison along with the modified expression. (From Harmer, S. W. et al., *IEEE Sensors Journal*, 99, 2013, doi: 10.1109/JSEN.2013.2273487.)

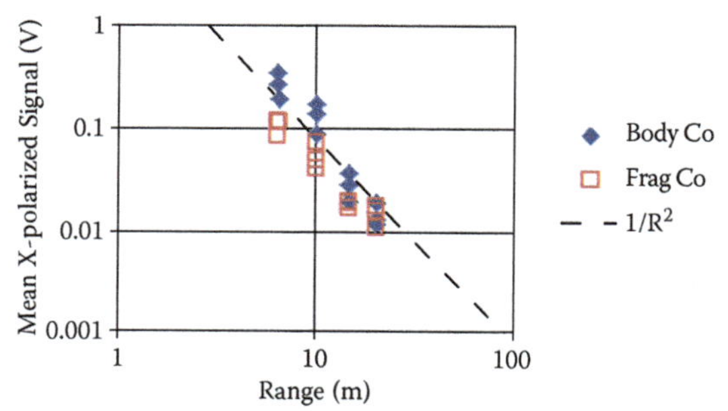

COLOR FIGURE 3.18

A log–log plot of the mean value of the measured co-polarized signal amplitude (V) from the human body as a function the range to the body. The body has no objects or has an attached PBIED. The inverse square range dependence is shown for comparison along with the modified expression. (From Harmer, S. W. et al., *IEEE Sensors Journal*, 99, 2013, doi: 10.1109/JSEN.2013.2273487.)

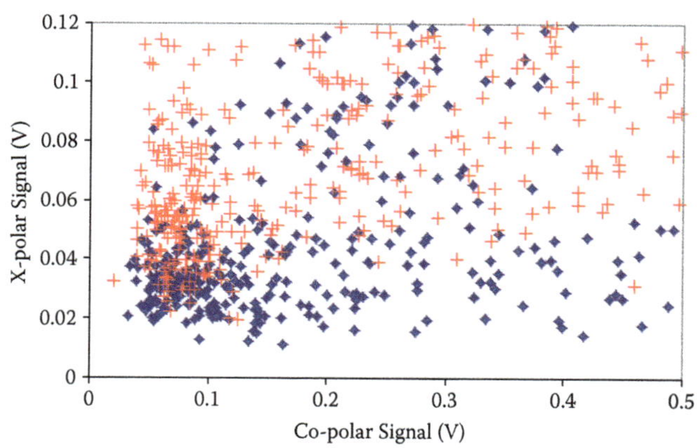

COLOR FIGURE 3.19

Measured co-polarized and cross-polarized signal amplitudes from 200 radar returns at a range of 6 m for a person with either a compact camera (diamonds) or small handgun (crosses) attached to the body. The groups of points are not clearly separated. (From Harmer, S. W. et al., *IEEE Sensors Journal*, 99, 2013, doi: 10.1109/JSEN.2013.2273487.)

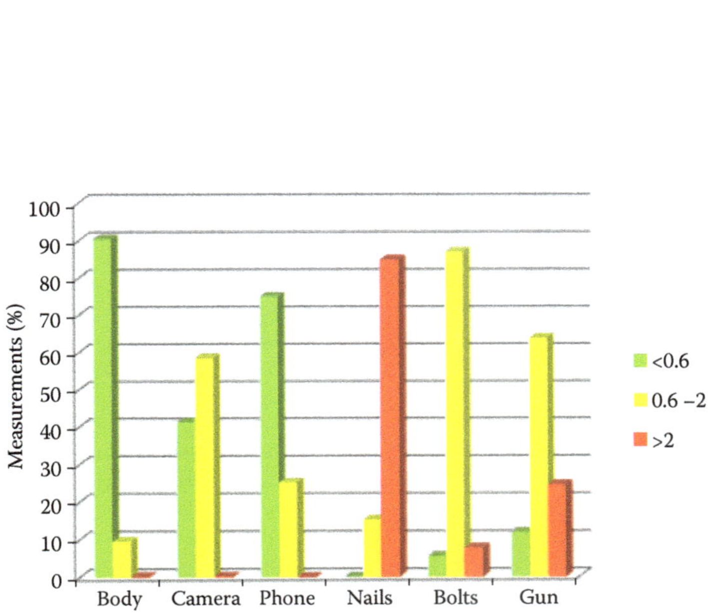

COLOR FIGURE 3.20

The percentage fractions of cross-polarized measurements which were below the lower threshold 0.6 (green), between 0.6 and 2 (yellow), and above 2 are shown for the different objects on the body. (From Harmer, S. W. et al., *IEEE Sensors Journal*, 99, 2013, doi: 10.1109/JSEN.2013.2273487)

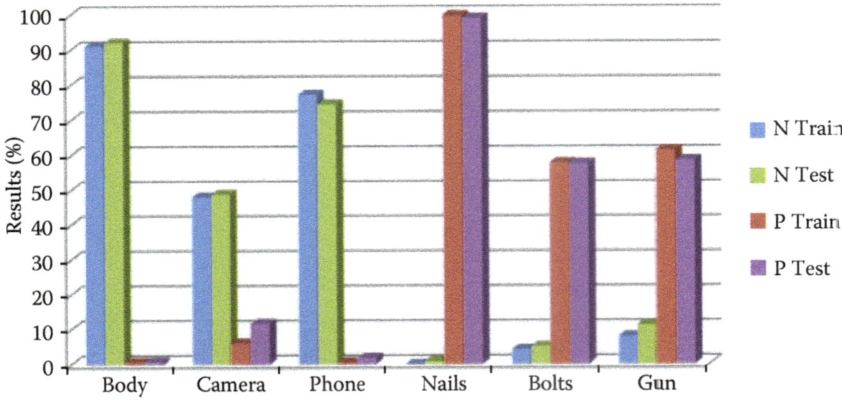

COLOR FIGURE 3.21

Sample results of neural network training using examples with all six object types. The height of the bars represents the percentage probability that the network assigns either N (negative threat) or P (positive threat), both for the set used in training the network and for the test set. (From Harmer, S. W. et al., *IEEE Sensors Journal*, 99, 2013, doi: 10.1109/JSEN.2013.2273487.)

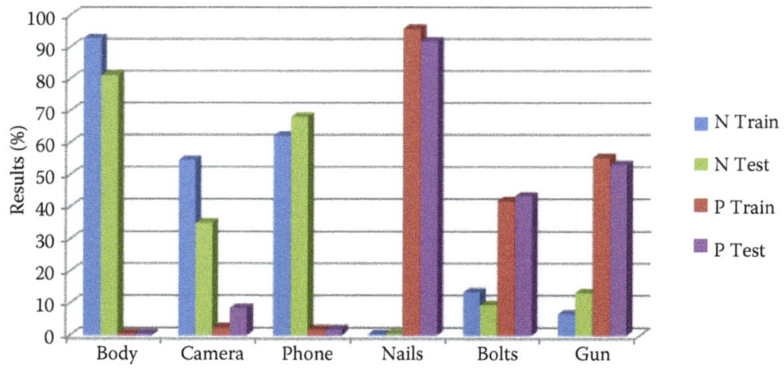

COLOR FIGURE 3.22

Examples of results from neural network training using a limited set of training objects. Results are shown as a percentage of classification results. (From Harmer, S. W. et al., *IEEE Sensors Journal*, 99, 2013, doi: 10.1109/JSEN.2013.2273487.)

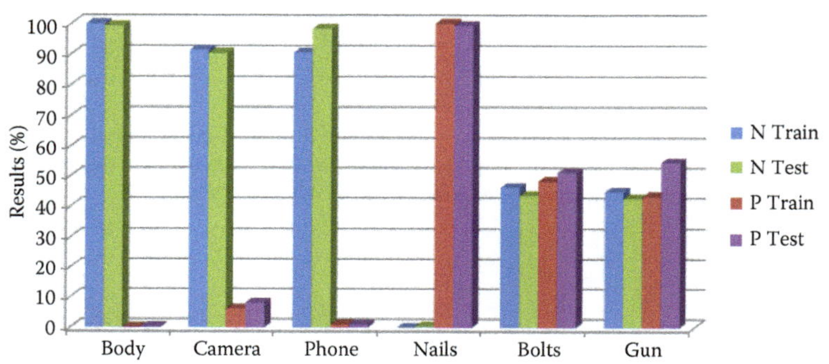

COLOR FIGURE 3.23

Examples of results from neural network training using only a PBIED with nails and the body alone in the training set. (From Harmer, S. W. et al., *IEEE Sensors Journal*, 99, 2013, doi: 10.1109/JSEN.2013.2273487.)

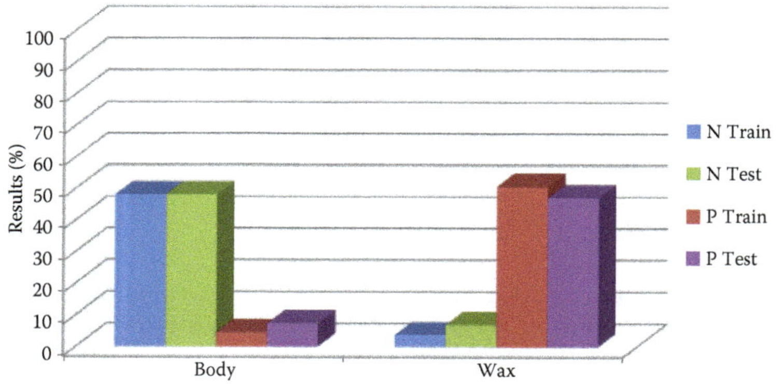

COLOR FIGURE 3.24

Sample results from neural network training, showing the discrimination of nonmetallic fragmentation PBIEDs. (From Harmer, S. W. et al., *IEEE Sensors Journal*, 99, 2013, doi: 10.1109/JSEN.2013.2273487.)

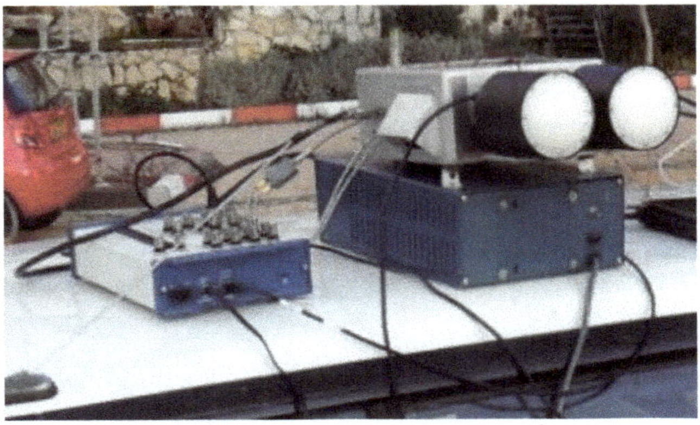

COLOR FIGURE 4.23
The modified sensor. (From Etinger, A. et al., *IEEE Sensors Journal*, 14, 4, 1232–1237, 2014.) © 2014 by the Institute of Electrical and Electronics Engineers [IEEE].)

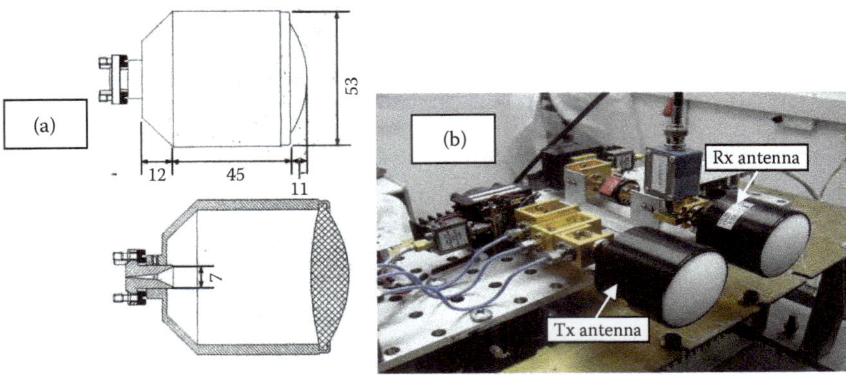

COLOR FIGURE 4.35
(a) The horn-lens Gaussian beam antenna and (b) general view of 330 GHz sensor with these antennae. (From Kapilevich, B. et al., *IEEE Conference on Microwaves, Communications, Antennas and Electronics Systems (COMCAS)*, 1–4, 2011. © 2011 by the Institute of Electrical and Electronics Engineers [IEEE].)

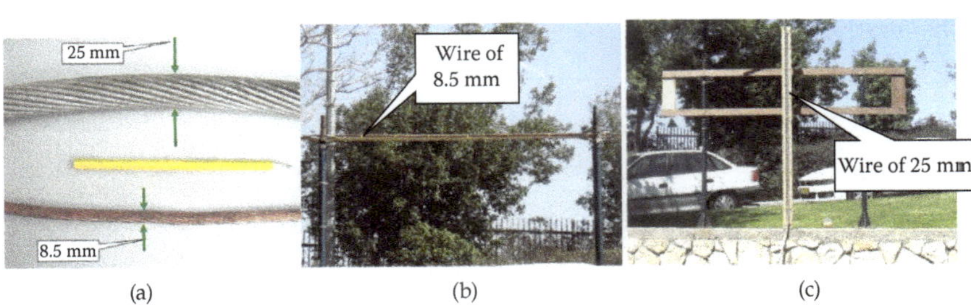

COLOR FIGURE 4.40
(a) The cooper line used in a 35 KV power network (diameter is 8.5 mm) and the aluminium line used in a 160 KV power network (diameter is 25 mm). (b) The section of copper wire mounted on a special stand in the horizontal position. (c) The section of aluminium wire mounted on a special stand in the vertical position.

COLOR FIGURE 6.11
Output signals in the outdoor scenario with two cars present.

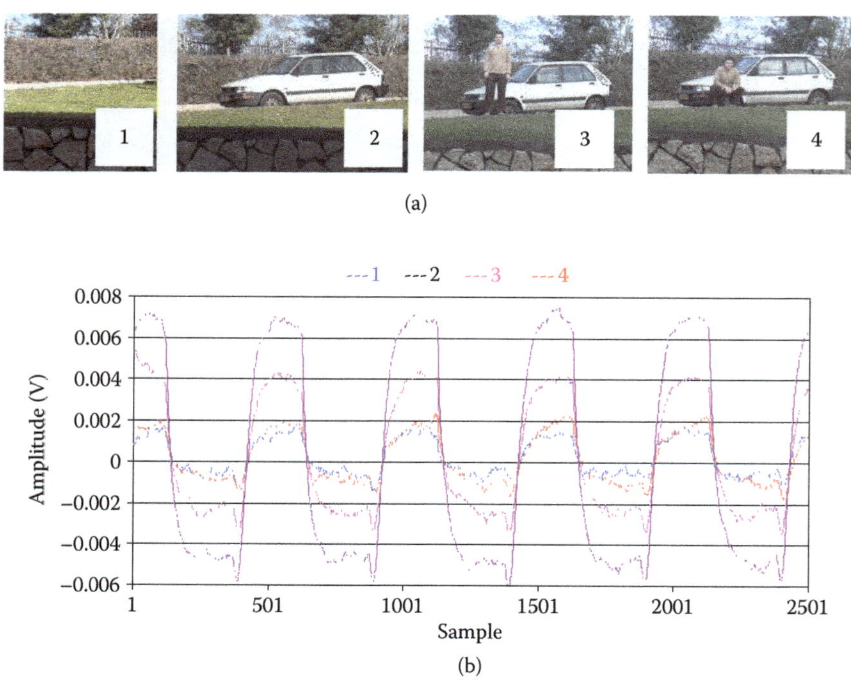

COLOR FIGURE 6.12
Mixed outdoor scenery: (a) a car and person and (b) recorded output signals. (From Kapilevich, B. et al., in *IEEE 24th Convention of Electrical and Electronics Engineers in Israel*, 161–164, 2006. © 2006 by the Institute of Electrical and Electronics Engineers [IEEE].)

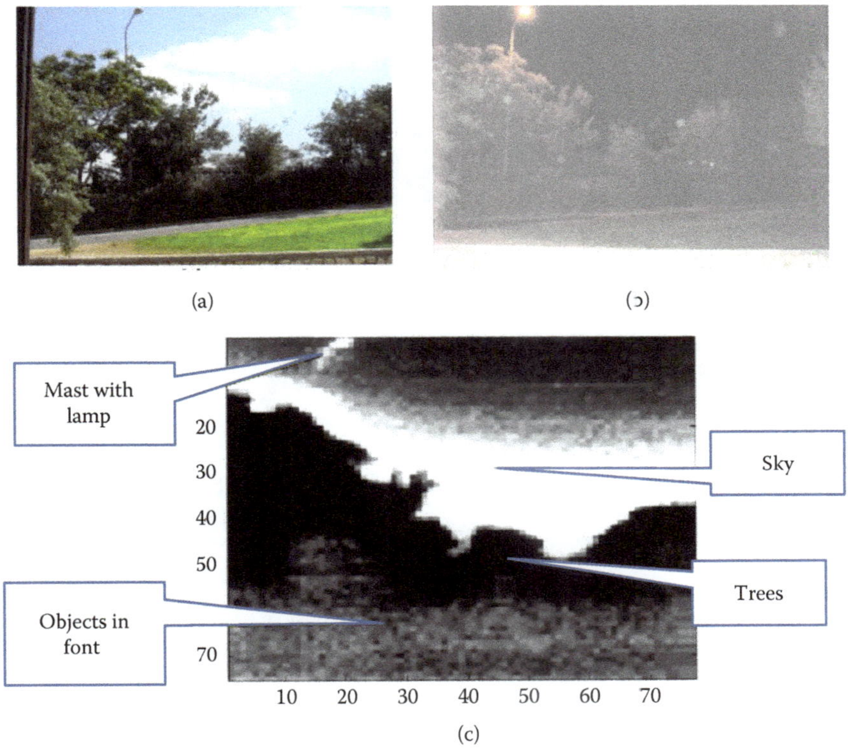

COLOR FIGURE 6.13
Mixed outdoor scenery: a streetlamp in the (a) day and (b) night and (c) the passive mm-wave image recorded at night.

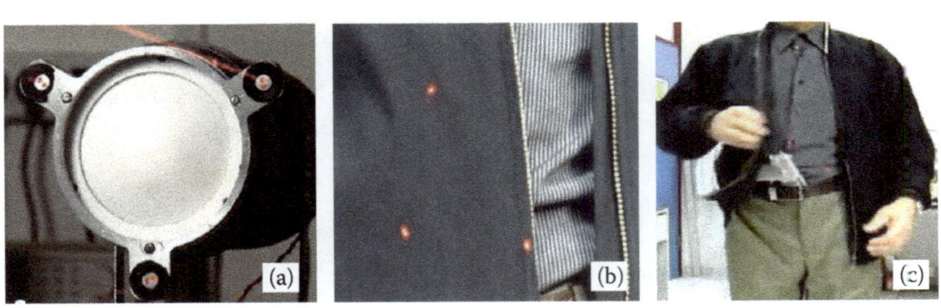

COLOR FIGURE 6.16
General picture of the sensor: (a) Teflon focusing lens and three laser pointers in front; (b) position of the beam spot during a scan indicated by laser pointers; (c) example of a concealed weapon. (From Kapilevich, B. et al., *IEEE Sensors Journal*, 13, 11, 4224–4228, 2013. © 2013 by the Institute of Electrical and Electronics Engineers [IEEE].)

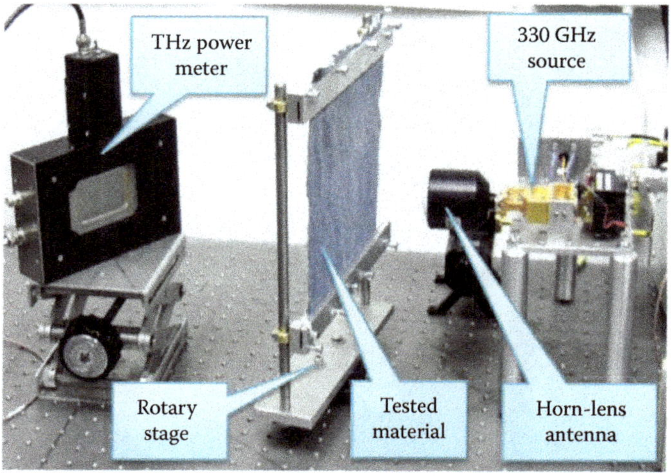

COLOR FIGURE 7.8
Photograph of the experimental setup. (From Kapilevich, B., Litvak, B., Anisimov, M., Hardon, D. and Pinhasi, Y., *International Journal of Microwave Science and Technology*, 2012, doi:10.1155/2012/375601. Reproduced courtesy of Hindawi Publishing Corp.)

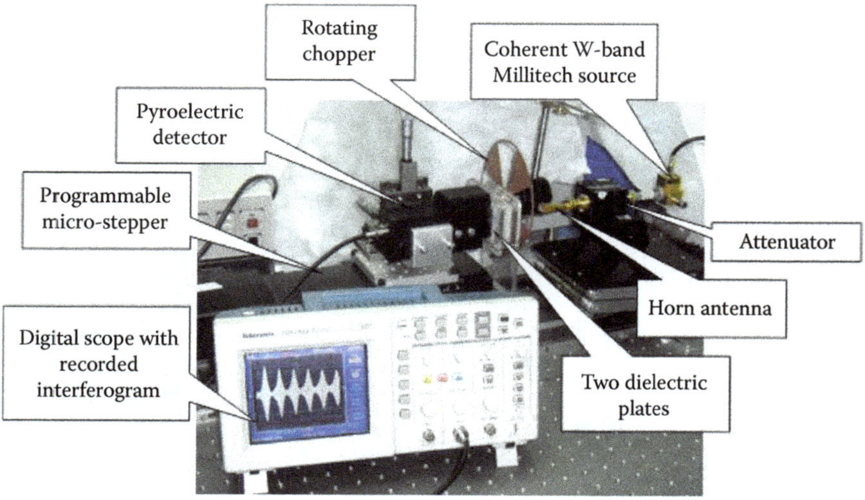

COLOR FIGURE 7.23
A view of the W-band experimental setup with a coherent source. (From Kapilevich, B. et al., *Progress in Electromagnetics Research B*, 28, 35–54, 2011. Reproduced courtesy of The Electromagnetics Academy.)

5

Active Microwave Sensors for Complex Natural Resonance-Based Object Detection

5.1 Introduction and Theory

Conventional magnetic portal screening systems cannot determine the number or type of metal objects concealed on the human body; instead, these devices merely indicate whether there is a significant metallic object or objects present [1]–[6]. In these detectors a nominally uniform magnetic field is applied within the immediate region of the portal; when the current generating this magnetic field is pulsed, metal objects have currents induced within them and the extended decay time of these eddy currents is used to detect the presence or absence of concealed metal objects. The human body, having a much lower conductance than metals, does not support eddy currents for the same timescales as exist with metal objects, and so the human body appears practically 'invisible' in these devices. The characteristic time periods for eddy current decay within handguns and other similar sized metallic objects are of the order of 100 μs [7], giving a characteristic frequency of the order of 10 KHz. At these frequencies, portable objects are electrically small and are located within the reactive near field of the antenna. In this regime, objects will not display any resonant behaviour, as would be the case with items commensurate in size with the wavelength of the electromagnetic excitation. This lack of resonant behaviour restricts the parameters available for object identification to a single parameter: the lifetime of the induced eddy current. Unfortunately, this is generally not sufficient to ensure robust discrimination of threat objects from benign objects, as there is considerable degeneracy in these lifetimes. Furthermore, the lifetime depends not only upon the shape of the object, but also on the metals present within the object. A different shaped and sized object composed of a different conductivity metal could thus mimic any metal object, for example, a stainless steel handgun [7].

A different approach, derived from an effect observed in ground-based radar systems for tracking aircraft, offers better discriminatory capability with multiple parameters which are independent of an object's aspect.

Complex natural resonances (CNRs) have been employed for many years to identify aircraft and missiles [8]–[19]. In order to excite the CNRs of typical handguns and similarly sized objects, the frequency of transmitted electromagnetic (EM) radiation must be ~500 MHz so that the wavelength is similar to the size of carried objects [20]–[24]. The CNRs for a sphere are well known and often referred to as the Mie solution of Maxwell's equations [25]. For a sphere much smaller than the incident wavelength, the radar cross section (RCS) is $\propto \lambda^{-4}$; in the Mie regime the RCS is oscillatory, and in the regime where the incident wavelength is much smaller than the sphere the RCS is constant and equal to the cross-sectional area of the sphere.

The scattered EM waves from a metal object illuminated with a broadband electromagnetic pulse contain aspect independent parameters, which depend only on object shape and the local dielectric environment [26]–[29]. The CNRs manifest in the temporal region, after the excitation pulse has passed the object, and this time signal is the late time response (LTR) of the object. Physically the LTR results from electromagnetic waves, which propagate around the object and scatter toward the receiver (see Figure 5.1).

Mathematically, the LTR can be expressed as a superposition of exponentially damped sinusoids [3], [8], [14], [15]. When the LTR of an object is sampled by a digital capture system, for example, a vector network analyzer (VNA), the signal h in the time domain data can be represented by

$$h[n] = \tfrac{1}{2}\sum_{m=1}^{M}\left(C_m \exp(Z_m n\Delta t) + C_m^* \exp\left(Z_m^* n\Delta t\right)\right) + N[n] \tag{5.1}$$

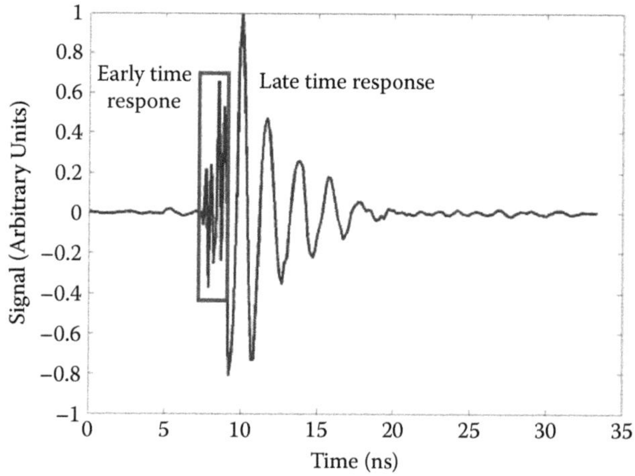

FIGURE 5.1
An example of a measured scattered return from a handgun positioned in an anechoic environment. The early time and late time response regions are marked.

Here $Z_m = -\alpha_m + i2\pi\nu_m$ are the aspect independent CNRs, where α_m and ν_m are the inverse lifetime and oscillating frequency, respectively. M is the model order (number of CNRs present in the LTR), and N is any signal that corrupts the LTR.

In general the model order, M, is unknown a priori and depends on the bandwidth of the microwave illumination, the shape and size of the object and the electromagnetic properties of the object's components and the local dielectric environment. Equation 5.1, excluding the noise term, is the impulse response of an object. Decomposition of this into a set of CNRs is the basis for concealed object identification. Complications in remotely measuring the impulse response of concealed objects arise for several reasons: antennae are required to transmit and receive electromagnetic signals. Each antenna has its own frequency response, and therefore filters the transmitted and received waveforms, corrupting the desired impulse response of a concealed object. Electrically small antennae cannot be used singularly, as, although such antennae present near-constant transmission and reception characteristics over a wide frequency band, the gain obtainable is too low for focused illumination. Larger antennae offer higher gain but present rapidly changing transmission and reception characteristics, and therefore seriously affect the ability to accurately measure an object's impulse function.

The antenna response for the transmitted and received signals needs to be corrected so that the LTR is associated with the target object rather than the antenna characteristics. Otherwise, CNRs of data will be corrupted and will include the CNRs of the antenna as well as those of any object. The measured signal will be a convolution of the transmitted excitation pulse, the impulse response of the object, the impulse response of the transmitter and that of the receiver.

$$S[n] = P[n] \otimes T[n] \otimes R[n] \otimes h[n] + A[n] \tag{5.2}$$

The symbol $\otimes$ denotes the discrete one-dimensional convolution operation, P is the pulse fed to the transmitter antenna, h is the target impulse response as given by Equation 5.1, and T and R denote transmitter and the receiver impulse responses, respectively. A is the received noise signal. The complex amplitudes C_m (known as residues) are dependent upon the aspect of the scattering object. The noise term encompasses any signals that do not have a foundation scattered from the object and which temporally overlap the LTR. This noise has a significant deleterious effect upon the ability to reconstruct the object's CNRs [20], [21].

Although an object will possess an infinite set of CNRs, we have observed experimentally that only the first few CNRs are excited with significant amplitude to be readily observable. It is also apparent that higher-order resonances tend to have shorter lifetimes, as they tend to be associated with smaller features of an object [22]. The selected model order need only be large enough to include lower-order modes, as it is these that dominate the LTR. By overestimating the model order, spurious CNRs are easily filtered by selecting only those CNRs with large amplitudes. This approach becomes less favourable when there is a large component of noise or signals which do not originate from the object's LTR. In these cases spurious CNRs with significant amplitudes are often present, and filtering by amplitude comparison is ineffective. In the case of considerable noise, for example, when the LTR contains reflections from the human body as well as from the concealed object, it is better to select a low model order. In practice, a model order of one is most suitable for detection and identification of concealed objects on the human body. With a model order of unity the LTR is decomposed into a single damped sinusoidal function; the intention is for this to be a good representation of the fundamental CNR of the object.

Use of this technique has been made in the field of radar airborne target identification where mathematical techniques are employed to decompose the LTR and find the CNRs of the object, which is usually an aircraft or missile. There are several mathematical processes proposed and used for CNR extraction from noise-corrupted data of the form given in Equation 5.1 [9], [10], [12], [18], [19], [30]–[32]. The generalized pencil of function (GPOF) method is often used and cited, as it is an efficient and accurate method which operates by solving a generalized eigenvalue problem [30]. The time domain LTR is decomposed into constituent CNRs which present as complex conjugate pairs ($Z_m = -\alpha_m + i2\pi\nu_m$ and its conjugate $Z_m^* = -\alpha_m - i2\pi\nu_m$); CNRs that present singularly are not physical, and this fact can be used to filter out spurious CNRs. Using the GPOF algorithm, the CNRs are recorded and the spurious CNRs are discarded. Other CNRs which clearly are not related to any concealed object are those with negative lifetimes, i.e. $\alpha_m < 0$; these are also discarded, as they give rise to exponentially growing signals which are not physical solutions [22].

5.2 Simulations with Electromagnetic Solver Software

Simulation of CNRs of objects with commercially available finite element software products, such as CST Microwave Studio, is a valuable tool. The prediction of an object's CNRs and their amplitudes allows a comparison to be made between predicted CNRs of an object and those observed experimentally. Based on these comparisons the CNRs of an object can be accurately estimated by careful simulation. Unlike a simple object, such as a cylinder or a sphere [25], [28], [33], [34], there are no analytical solutions for the CNRs of complicated objects like a handgun. Therefore, modelling becomes important as one considers the frequency range in which illumination must occur

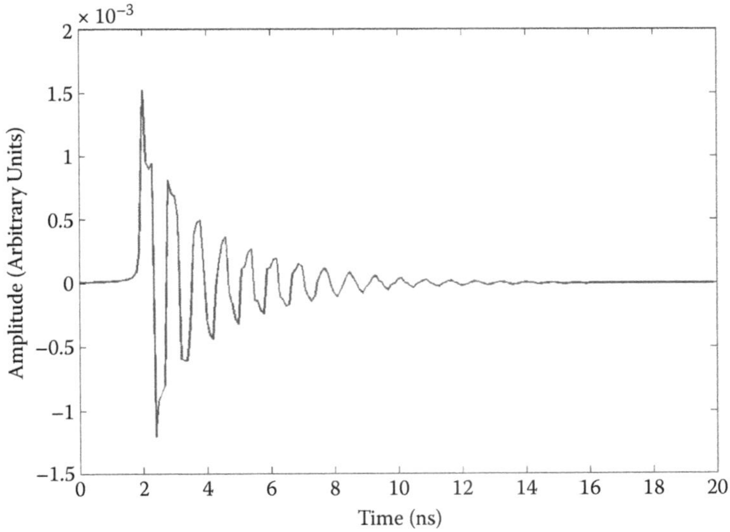

FIGURE 5.2
The approximate impulse response of a 10 cm length of wire as simulated in CST.

for excitation of the fundamental CNRs of a range of objects and the size and directional pattern of the radar cross section (RCS) exhibited.

Figure 5.2 shows the impulse response of a 10 cm length of perfect conducting wire with 2 mm diameter. By simulation in CST Microwave Studio, the wire is illuminated with a plane wave having a near-uniform frequency power spectrum over the 1–10 GHz range. 1001 frequency points are used in the simulation. The number of mesh cells required to adequately simulate the electromagnetic interaction with the object is ~40,000, and the accuracy of the transient solver is set to –80 dB. The frequency domain data from the simulation were processed in MATLAB® by performing an inverse fast Fourier transform (IFFT) to transform to the time domain, using the pulse synthesis method [35]. Using experimental data taken in the laboratory, using the apparatus and methods discussed in Section 5.3, a comparison can be made between the simulated and experimental LTR of the 10 cm thin-wire object (see Figure 5.3).

The GPOF algorithm [30] implemented in MATLAB® is used to estimate the CNRs by decomposing the LTR. Applying the GPOF algorithm to the LTR data for the simulated 10 cm wire, with a model order set to two, gives CNRs of 0.453 + *i*1.44 and 0.5013 + *i*4.39. The first CNR corresponds to a fundamental resonant frequency of 1.44 GHz and an exponential decay rate (α) of 0.453 ns^{-1}, equivalent to a lifetime of 2.21 ns. The second CNR is clearly the third harmonic of the fundamental frequency. This is expected, as the electromagnetic field distribution must be symmetrically distributed along the

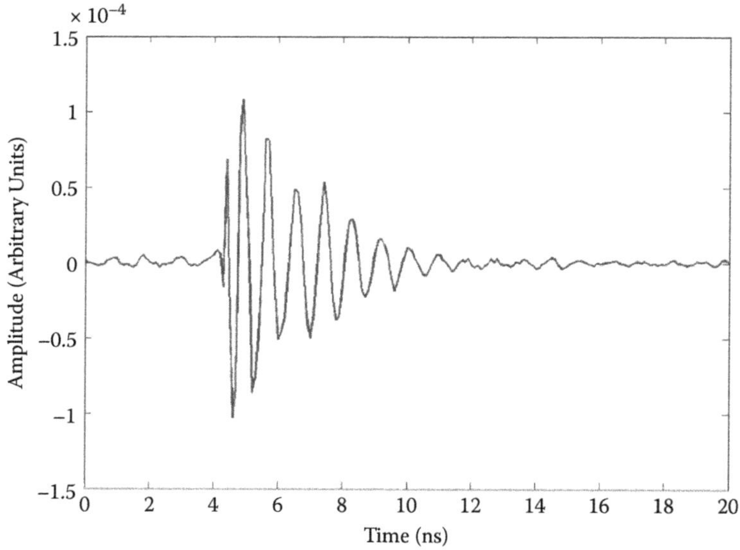

FIGURE 5.3
The approximate impulse response of a 10 cm length of electrical wire as measured in the laboratory.

length of the wire, permitting only odd harmonics of the fundamental. The value of α is expected to remain constant as the CNRs are only associated with the length of the wire [36]. For more complicated shapes, higher-order CNRs are associated with different structures, for example, the trigger and guard of a handgun, and in these cases the higher-order CNRs may have frequencies which are not associated with the fundamental [22].

The fundamental CNRs of the wire, obtained from measurements made in the laboratory, were found to be $0.621 + i1.39$ and $0.783 + i4.28$. The simulated and measured LTRs present similar appearances (see Figures 5.2 and 5.3).

A selection of five representative objects—an automatic handgun, a revolver handgun, a knife, a pen and a mobile phone—were modelled and run in CST to find their impulse responses. The CNRs were then found by application of the GPOF algorithm to the computed LTR. These CNRs are listed in Table 5.1. The limited bandwidth (0.3–6 GHz, stepped over 1001 points) of the pulse means that only a small set of CNRs are presented for each object. It is clear that some objects present a greater number of CNRs than others; for example, the knife presents only two clear CNRs in this frequency range, while the mobile phone displays seven. Each of the objects has a fundamental CNR which is different from that of any of the other objects (see Figure 5.4) and may be used as the basis for object identification.

Simulations of the far-field radar cross sections (RCSs) of objects were also made in CST; these simulations are useful, as they show the complicated

TABLE 5.1

Common Objects and Their CNRs

Object	CNR ($-\alpha$ ns^{-1} and ν_m GHz)	Model
Mobile Phone Handset	1.37, 0.961 0.344, 1.58 0.922, 2.07 0.815, 2.81 0.892, 3.75 0.368, 4.43 1.53, 4.95	
Fountain Pen	1.05, 1.03 0.139, 3.13 0.376, 5.93 2.54, 3.24 2.16, 2.15	
Handgun Revolver	0.714 , 0.644 1.14 , 1.24 0.369, 2.16 1.30, 3.51 2.21, 4.06 2.31, 6.92	
Handgun Glock	0.92, 0.48 1.43, 0.81 0.93, 1.95	
Knife		

directivity radiation patterns that most objects produce and the overall RCSs of objects in the range of their fundamental CNR frequencies. The RCS patterns for an automatic (Glock) handgun and a knife are calculated for a cross-polarized configuration (see Figure 5.5). Cross-polarization is useful, as it reduces the power reflected from a human body, giving greater ability to discriminate most metallic objects concealed on the body. See Section 5.3 for

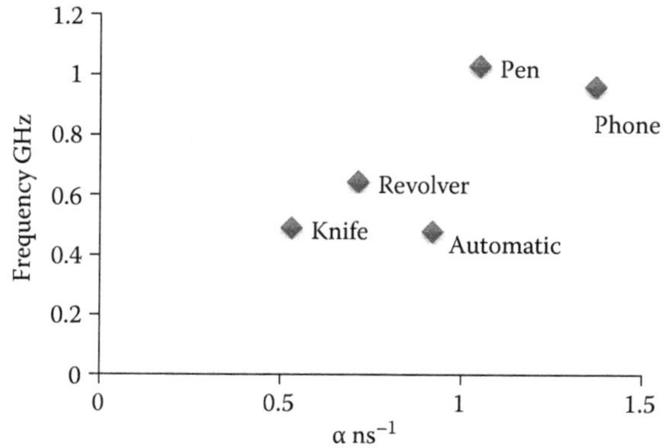

FIGURE 5.4
Graphical plot showing the fundamental complex natural resonant frequencies of the objects detailed in Table 5.1.

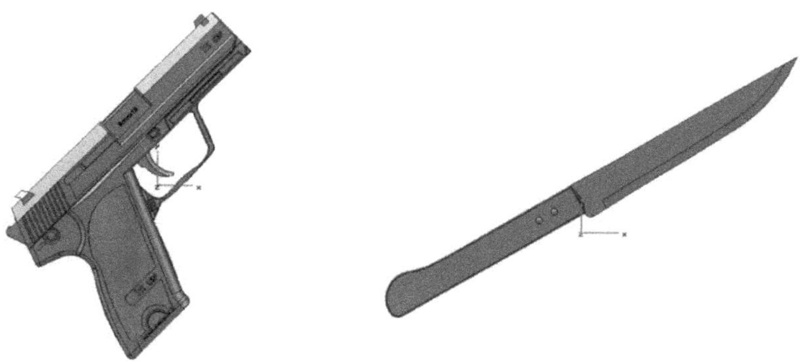

FIGURE 5.5
Models of the automatic (Glock) handgun and the knife used for RCS calculation.

further details. The weapons were illuminated side on, as shown in Figure 5.5, the RCS results were are plotted as a function of angle (see Figure 5.6).

The averaged RCSs over the frequency range 0.3–6 GHz for the automatic handgun and the knife are shown in Figure 5.6. The knife has a far narrower RCS pattern than does the gun; due to its flat-sided nature, it presents a large RCS when side on and virtually zero RCS when head on. The knife also exhibits a more symmetrical RCS pattern, while the RCS pattern of the handgun is broader, less symmetric and has a lower peak value. From these plots, it is readily apparent that the detection of a knife may prove more difficult than for a handgun due to a narrower RCS pattern, giving the knife a large RCS only in a narrow range of orientations.

RCS plots at single frequencies, within the 0.3–6 GHz range, show how rapidly the angular distribution of the RCSs of the handgun and knife change with frequency (see Figure 5.7). At lower frequency, the handgun gives broad

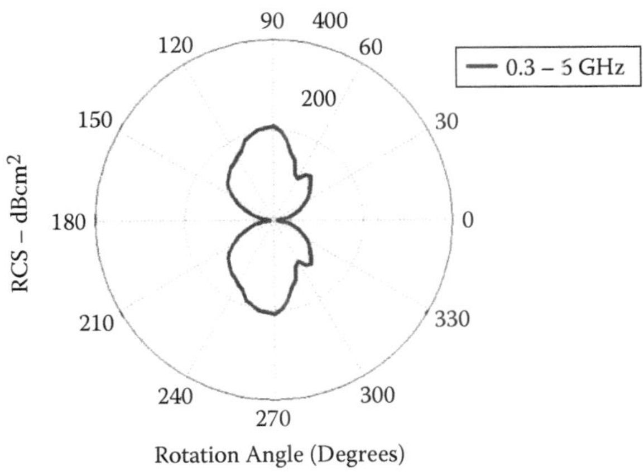

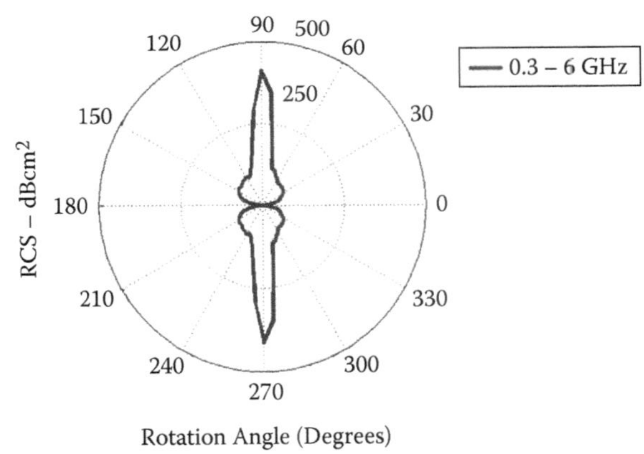

FIGURE 5.6
The averaged cross-polarized RCS over 0.3–6 GHz of a 9 mm Glock handgun (top) and knife (bottom) as a function of orientation with respect to a horizontally polarized electric field. The incident plane wave propagates from the 90° point toward the 270° point. Unsurprisingly, both objects exhibit maximal RCS when side on (90° and 270°) and minimum RCS when head on (0° and 180°).

lobes in its RCS pattern. When the fundamental resonance of the handgun is excited, at 0.48 GHz, the RCS becomes significantly larger, indicating that the incident radiation is being coupled into waves propagating around the object, which serve to increase the RCS. This effect is commonly discussed in the Mie solution for a sphere [25]. After the fundamental CNRs, the RCS falls off, but the RCS angular pattern becomes more complex as the ratio of the wavelength to object size increases, giving a more specular reflection. At higher frequencies corresponding to higher-order CNRs (0.94 and 1.85 GHz) the RCS again increases. The knife shows a similar pattern with an increased

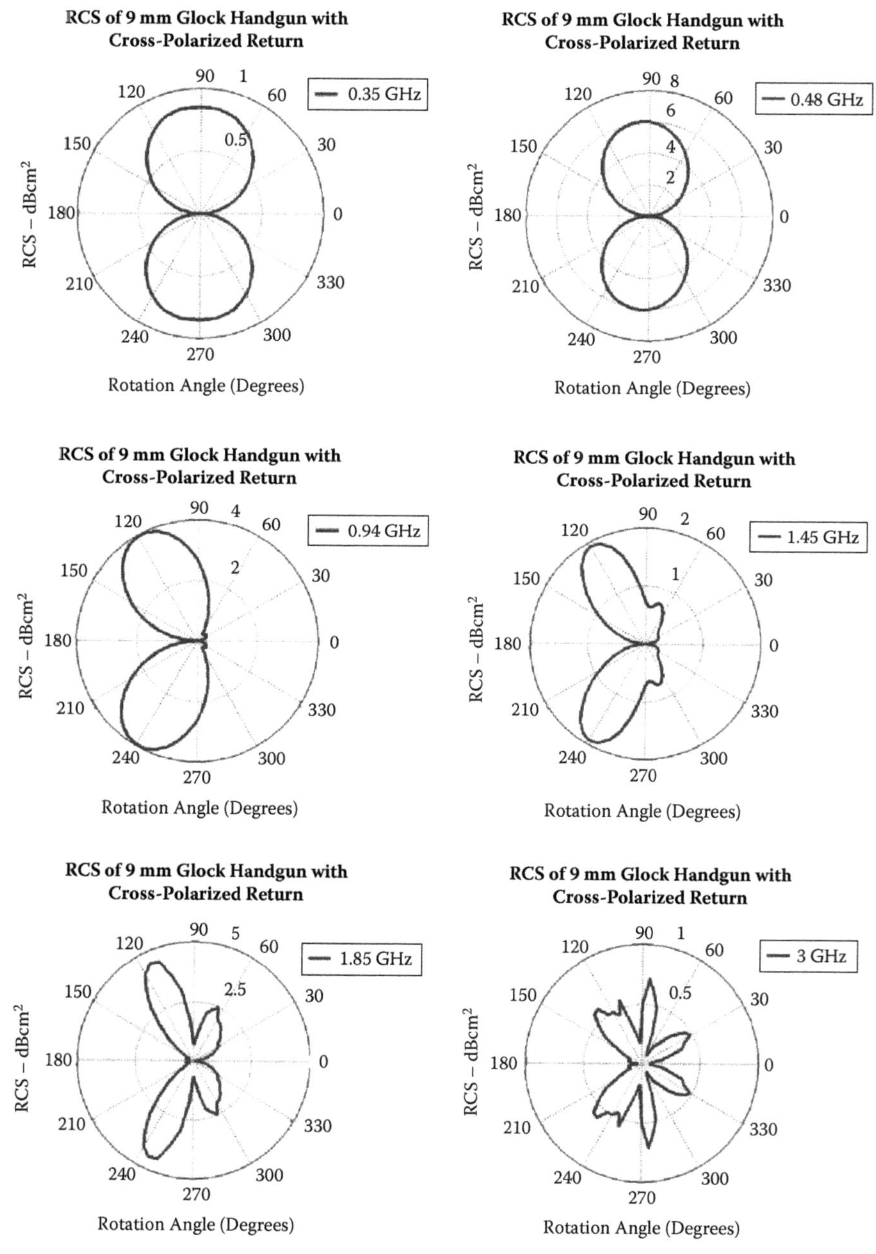

FIGURE 5.7
The RCS patterns for a knife and a handgun at selected frequency points around the fundamental CNR frequencies for both objects. *(Continued)*

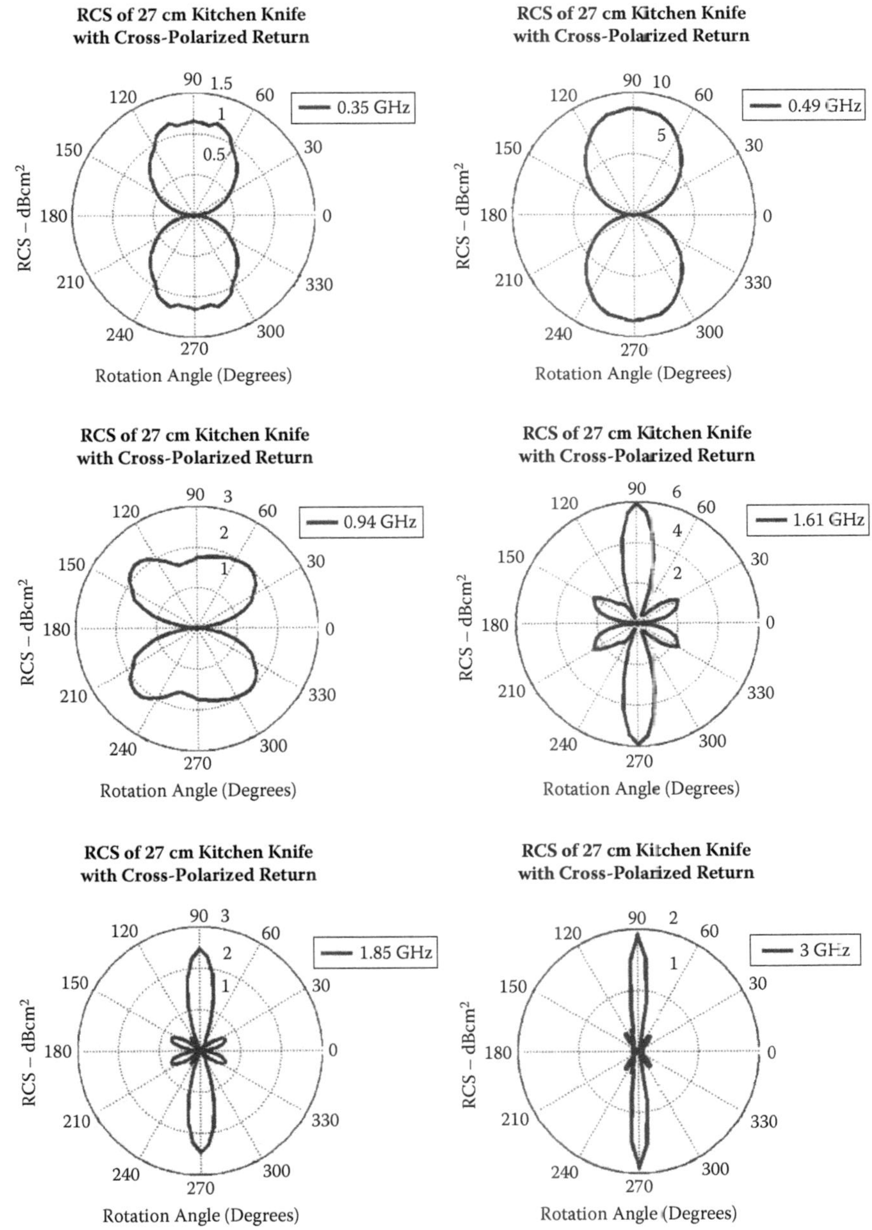

FIGURE 5.7 (Continued)
The RCS patterns for a knife and a handgun at selected frequency points around the fundamental CNR frequencies for both objects.

RCS at its fundamental CNR, but the RCS angular pattern is more symmetric and narrower than is the case for the handgun.

5.3 Experimental Configuration, Signal Processing and Results

An Agilent E8363B vector network analyzer with waveguide horns attached to ports 1 and 2 was used to transmit and receive ultra-wide band (UWB) microwave signals by stepped frequency transmission (see Figure 5.8 and Figure 5.9). Both horns, in a bistatic arrangement, are directed at the target, and the amplitude and phase of the reflected field recorded at 1024 equally spaced frequencies over the band 0.4–6 GHz. Data were taken inside of an anechoic chamber to minimize the deleterious effects of background clutter on the LTR of any object. In cases where only an isolated object is measured, the object was suspended by string inside the chamber; for on-body measurements a volunteer stood, nominally motionless, within the chamber with any object required concealed on the front of the abdomen.

The human body and any background objects scatter any microwaves used for person screening; this effect provides the major difficulty in implementing a reliable screening system using the CNRs of objects [21]. The effect of noise from other scattering objects on faithfully reconstructing the impulse function of a concealed object is detailed in Equations 5.4–5.6. The effect can

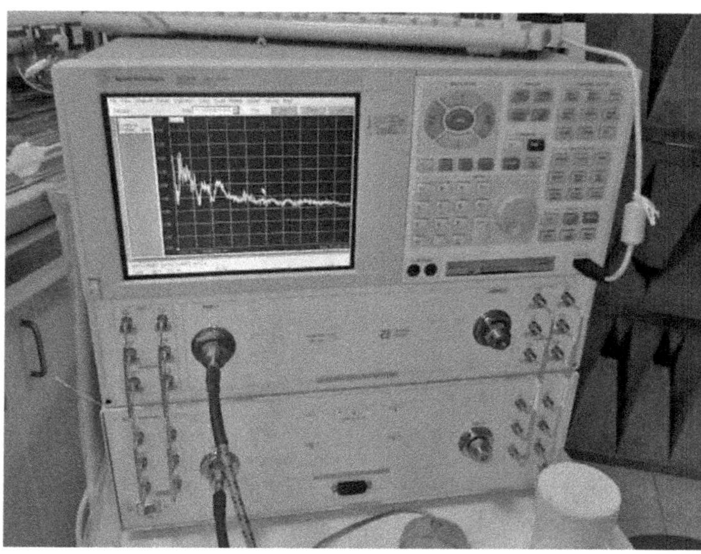

FIGURE 5.8
Agilent VNA (0.1–40 GHz) used in experimental investigations into CNR-based concealed object detection on the human body. Typically two ports were used, one for transmission and another for reception.

FIGURE 5.9
Pairs of Vivaldi horn antennae used to excite and measure the CNRs of objects concealed on the body. The smaller antennae (left) operate over the frequency band of 1–18 GHz, while the larger pair (right) operates over 0.4–3 GHz.

be reduced, to some degree, by making use of the fact that the human body scatters incident EM radiation predominantly in the same plane of polarization as the illuminating radiation, while complex-shaped metal objects such as handguns scatter strongly in plane and out of plane [20]. Therefore, an orthogonally polarized transmitter and receiver (see Figure 5.10) can improve the signal to the clutter level significantly [37], [38].

For laboratory testing, both in isolation and on the human body, a selection of typical concealed weapons is used. There are three different handguns: an air pistol, a 5.56 mm Olympic style starter revolver and a replica 9 mm Glock handgun. The weapons are similar in size and shape to live firing handguns. The kitchen knives are of a common design and consist of a stainless steel blade encased within a plastic handle.

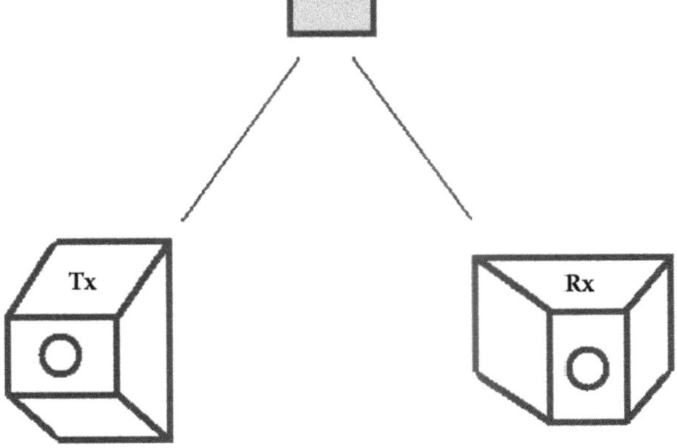

FIGURE 5.10
Illustration of orthogonal arrangement of antenna that permits increased discrimination of the LTR from reflections from the human body.

FIGURE 5.11
A selection of weapons used for measurements of CNRs in isolation and on the human body.

The weapons displayed in Figure 5.11 were used individually or on the human body within an anechoic chamber. The larger, 0.4–3 GHz, horn antennae (see Figure 5.9) were used in an off-axis, bistatic radar setup. Illumination was with the electric field horizontally polarized; the receiving antenna was vertically polarized to reduce the effects of background clutter. The antennae were driven by the VNA over a range of 0.4–6 GHz at a power level of 0 dBm. 6 GHz was selected at the upper frequency limit, as this is where the antenna voltage standing wave ratio (VSWR) exceeds 1:2.

The GPOF algorithm has been used to calculate the CNRs of the retrieved impulse response of all the weapons used in the laboratory experiments. The fundamental CNRs for the 9 mm Glock handgun, air pistol and 26 cm kitchen knife are not dissimilar, and therefore higher-order CNRs may be required to differentiate between these objects.

The CNRs listed in Table 5.2 are essential when working experimentally with objects on the human body. It may be supposed that an object with a particular fundamental CNR will increase the cross-polarized RCS of a person carrying that object. However, it must be noted that the local dielectric environment will affect the CNR of an object. For example, the CNRs of unexploded ordnance (e.g. artillery shells) are shifted from their position in free space when buried in sand [26], [36], [39]. A similar but greater shift is expected from any object positioned on or near the human body; the shift is likely to be larger than for low-loss low-dielectric-constant materials like sand because the human body has a very high dielectric constant with large loss tangent at microwave frequencies [40]–[42].

TABLE 5.2

CNRs Obtained from the Measurement of Various Isolated Weapons in an Anechoic Chamber

CNR	9 mm Glock Handgun	Olympic Starter Pistol	Air Pistol	13 cm Kitchen Knife	26 cm Kitchen Knife
1 (Fundamental)	0.48 + *i*0.51i	0.59+*i* 0.61	0.41 + *i*0.48i	1.18 + *i*1.24	0.49 + *i*0.53
2	1.07 + *i*0.88	1.24 + *i*1.19	1.11 +*i*0.99	1.55 + *i*2.08	0.99 + *i*1.61
3	0.92 + *i*1.55	0.98 + *i*1.89	0.97 + *i*1.67	—	—
4	1.64 + *i*2.53	0.71 + *i*2.13	—	—	—
5	—	1.54 + *i*2.99	—	—	—

A further problem that arises when the CNRs are to be used for object identification is the separation of the reflected incident wave, or early time response, from the LTR. The start of the LTR is ill-defined [43], but a perfectly suitable definition that is used is

$$t > T_R + T_O + T_P \tag{5.3}$$

where T_R is the round-trip time from antenna to object, T_O is the transit time across the object, and T_p is the temporal width of the transmitted pulse. If pulse synthesis is used, for example, using stepped frequency to synthesize a pulse, then the temporal width of the synthesized pulse is used.

Any signal returning after this time must be the result of multiple reflections within the object, for example, creeping waves that have circumnavigated the object or waves having suffered successive reflections inside of a transparent dielectric object (see Figure 5.12). However, signals which suffer multi-path scattering may not be associated with the LTR of the concealed object, but due to their elongated path, may temporally overlap the LTR and therefore corrupt this signal.

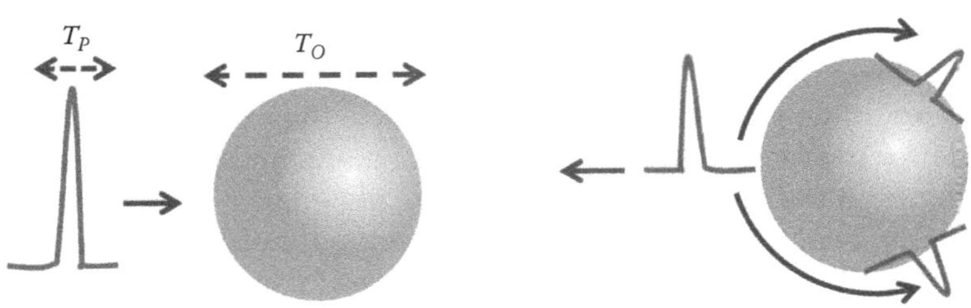

FIGURE 5.12
An illustration of the origin of the LTR for a sphere: The incident wave (top left) is scattered from the surface of the sphere and electromagnetic waves propagate around the surface of the sphere (top right). The incident pulse transits the sphere, while the surface waves continue to propagate (bottom left). Radiation is emitted back toward the receiver with a time delay after the initial reflection (bottom right).

Removal of multi-path scattering is impossible in all but the most controlled of environments, for example, when the object is located in an anechoic environment. However, the effect of multi-path signals can be mitigated to a certain extent by directing the incident microwave radiation as tightly as possible on the portion of the body being screened and by using an orthogonally polarized transmitter and receiver to further improve discrimination [20]. Due to the wavelength required to excite the fundamental CNRs of typical handgun-sized objects (~20 cm), a very large aperture is required for effective focusing. Such focusing may be implemented in the near field using a synthetic aperture such as a phased array [21].

Frequency spectrum data taken from the VNA are processed in MATLAB®, where a background subtraction method is utilized. This procedure is essential and cancels the effects of direct cross talk between the transmitter and receiver antenna. The background data are taken without any object or person present in the zone of the antenna. Background subtraction works extremely well in static situations; however, when there is a change in the background scenario, such as when a large target is being screened (shielding a significant portion of the background), or where objects in the background are moving, the procedure is ineffective. Although highly absorbing, the radar absorbing material used in the anechoic chamber in which experiments were conducted does reflect a small fraction of microwave radiation. This radiation, together with the direct cross talk between antennae, constitutes the background signal.

Removal of the initial reflection from the received signal is required to isolate the LTR, which can then be operated upon to give an object's CNR. Empirically, for objects on the body, we have determined that by extracting ~10 ns of time domain data immediately after the maximum reflected signal (which usually corresponds to the initial reflection from an object), the LTR can be well isolated from the higher-frequency initial reflection.

Before the CNR can be accurately estimated, the antenna response should be removed from the resulting LTR by a process of deconvolution.

It is unfortunate that suitable Vivaldi antennae for CNR excitation of concealable objects such as handguns, knives, etc., are similar in size to these objects, and thus present with similar CNRs of their own (see Figures 5.13 and 5.14 for measured CNR). The CNRs of an antenna can be imagined as resulting from radiation reflecting from any impedance mismatches within the antenna and antenna feed and being emitted with some delay relative to the main waveform in much the same way as for creeping waves around a sphere (see Figure 5.12).

A conventional method to measure antenna response is to align two antennae in the far-field region. A transmission power of 0 dBmw was used, and the frequency range of 1–10 GHz was divided into 801 frequency points. With the smaller, 1–18 GHz Vivaldi horn antennae (see Figure 5.9), the results of the direct transmission measurement are shown in Figure 5.15. The sensi-

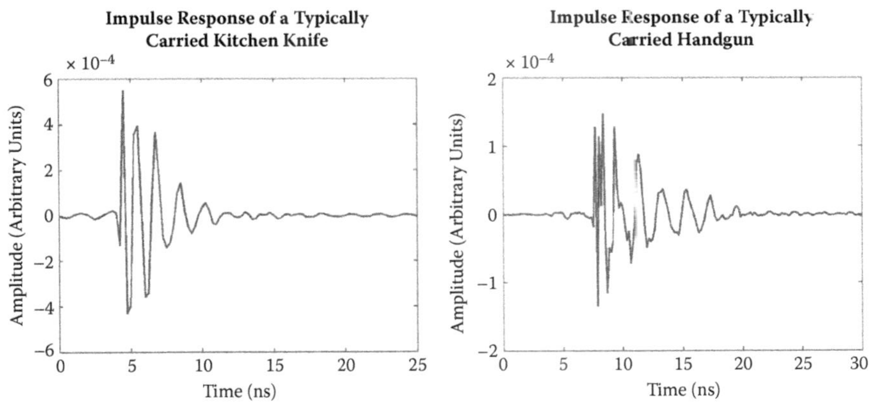

FIGURE 5.13
Impulse responses obtained in the laboratory for a knife and a handgun in isolation.

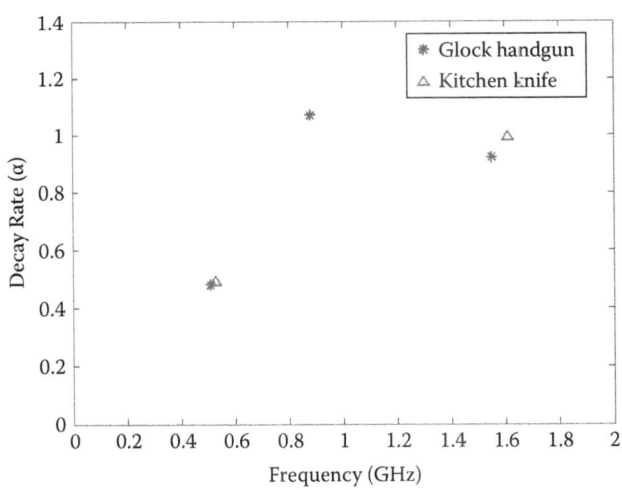

FIGURE 5.14
A plot to show the CNRs obtained for the 9 mm Glock handgun and the 26 cm kitchen knife when illuminated by a UWB 0.4–6 GHz pulse.

tivity of the antenna response on alignment angle is shown by aligning the antenna slightly off-axis.

Figure 5.15 shows that the antenna is more directive, having a narrower beamwidth, at higher frequencies. The angular dependence of directivity makes robust correction for the antenna response possible only in a single, chosen direction. However, since the lower-frequency directivity is far less angularly dependent, the effect of deconvolution with a single antenna response is greatest on the initial reflection and less influential around typical CNR frequencies. A further problem is the relatively low-transmission

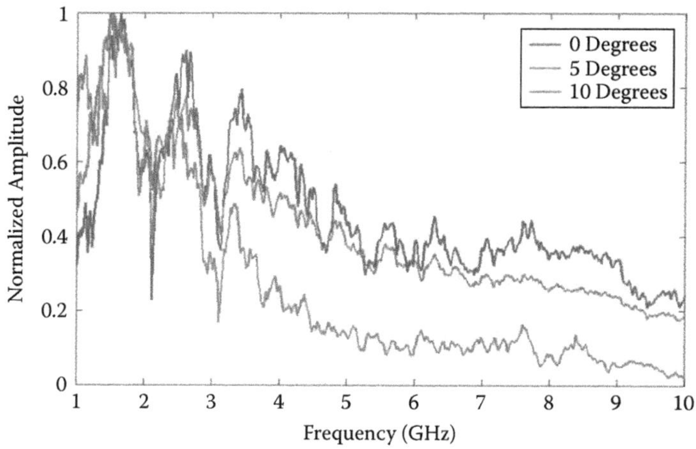

FIGURE 5.15
Vivaldi horn antenna response for direct transmission over 1–10 GHz at three off-axis angles.

powers in specific frequency ranges, for example, at ~2.1 and 3.1 GHz. Deconvolution by division of the received signal by the antenna response in the frequency domain can result in anomalous CNRs at regions where the transmitted power is low.

Recasting Equation 5.2 in the frequency domain by applying the convolution theorem, the equation becomes

$$\tilde{S}[m] = \tilde{P}[m]\tilde{T}[m]\tilde{R}[m]\tilde{h}[m] + \tilde{A}[m] \quad (5.4)$$

where the tilde denotes the amplitude in the frequency domain and m labels the frequency point. The noise term, $\tilde{A}$, includes any signal which is not scattered from a concealed object; this can be reflections from the human body and other objects in the vicinity of the person being screened. Even in an anechoic environment, the reflection from the human body is easily sufficient to corrupt the LTR of any concealed objects and make extraction of their CNRs a challenging task. Unfortunately, the form of this noise is unpredictable. For example, the scattering depends very strongly on the size and orientation of the human body in a way which cannot be reliably predicted. Removal of the antenna response can artificially increase this effect at certain frequencies where the antenna response is low.

If the receiving and transmission antenna are nominally identical, then

$$\tilde{S}[m] = \tilde{P}[m]\tilde{T}^2[m]\tilde{h}[m] + \tilde{A}[m] \quad (5.5)$$

By measuring the product $\tilde{P}[m]\tilde{T}^2[m]$ (see Figure 5.15), the antenna response can be corrected for in the frequency domain before application of IFFT to give the time domain signal from which the LTR can be isolated from the initial reflection. Thus,

$$\frac{\tilde{S}[m]}{\tilde{P}[m]\tilde{T}^2[m]} = \tilde{h}[m] + \frac{\tilde{A}[m]}{\tilde{P}[m]\tilde{T}^2[m]} \tag{5.6}$$

The left-hand side of Equation 5.5 is the deconvolved response and is subject to an IFFT to give the scattered time domain signal. The time domain signal is then separated into the initial reflection and LTR by time gating of the temporal response. This process should yield an accurate reproduction of the impulse response of any concealed object. However, the noise term is also included, and this term can easily be larger in amplitude than the impulse response that is being sought. The noise term can be amplified considerably by the deconvolution procedure at frequency ranges where the antenna response is low. This effect seriously degrades the recovered impulse response, as the noise term is generally present at all frequency ranges transmitted. Typical coupling around a human body gives a path length of ~1 m, equating to a transit time of ~3 ns. The lifetime of the fundamental CNR of a handgun is ~1–2 ns; hence, the LTR of the human body will certainly temporally overlap the LTR of a concealed handgun. Furthermore, the amplitude of the LTR from the human body can be many times larger than that from a concealed object and the form of the body's LTR is complicated, and highly variable, making separation of the two signals a near impossible task in most practicable environments.

The GPOF method for decomposing an impulse response into CNR frequencies is an excellent method when the noise term is small and that noise is Gaussian distributed in time [44]. However, in a realistic detection scenario these conditions are not true; the noise is often larger than the signature being sought, and the noise is certainly not statistically similar to random processes like thermal or photon noise. Although the GPOF method is very useful for ascertaining the CNRs of objects in isolation, by processing the impulse response of simulated or actual objects measured in the laboratory, the GPOF method is of limited use for on-body detection of concealed objects.

A time–frequency approach [45] can provide graphical output, which is readily interpretable and may be used to show the presence of CNRs associated with objects concealed upon the body, and as a result is less susceptible to corruption by reflection from the body and other background clutter. The visual interpretation of the LTR using the time–frequency approach shows the frequencies present for a portion of the LTR length. Implementation is by short-time Fourier transform (STFT) of the scattered signal with a window function which serves to select only a fixed temporal width of the signal at any one instant. By sliding this window over the time signal, the frequency spectrum of the LTR and the higher-frequency initial reflection are also easily separated.

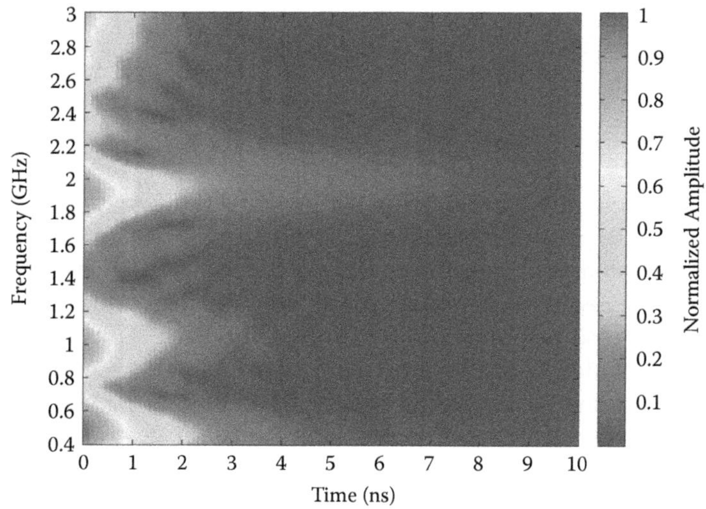

FIGURE 5.16
Time–frequency plot of a 9 mm Glock handgun; four distinct CNRs are clearly visible.

STFTs with fixed time span windows of 5 ns are applied to 10 ns of LTR data. The window is moved through the LTR by shifting the window position one data point for each STFT. The data points are separated by the time interval $1/BW$, where BW is the total transmitted bandwidth. TF analysis from a 9 mm Glock handgun measured in an anechoic chamber is shown in Figure 5.16. The handgun was aligned with a barrel perpendicular to the incident 0.4–6 GHz UWB illuminating pulse. The frequency data obtained were subject to an IFFT to transform to the time domain. The LTR was extracted from the maximum absolute amplitude point using the procedure described earlier.

The CNR of the Glock handgun can be clearly identified using the TF analysis method (see Figure 5.16). The four CNRs are seen to occur at 0.48, 0.97, 1.85 and 3.01 GHz. The lifetimes, however, are more difficult to determine accurately using the TF method. The frequency width of the peaks of the CNR lifetimes are related to their lifetime in such a way that longer-lived resonances will be sharper, while shorter resonances will present a broader peak.

The TF plot for a human without any objects is shown in Figure 5.17. There is a considerable LTR resulting from EM waves being diffracted around the body. The lifetimes of ~2–3 ns coincide with the time taken for the waves to circumnavigate the body. The frequency bands which these LTRs occupy depend on the position, stance and size of the human body and are not easily predicted. When the person has a handgun concealed on the abdominal area, the TF plot shows a distinct alteration of the pattern of the frequency positions, lifetimes and amplitude of captured LTRs (see Figure 5.18). The first two CNRs of the handgun at 0.92 ns^{-1}, 0.48 GHz and 1.43 ns^{-1}, 0.81 GHz are

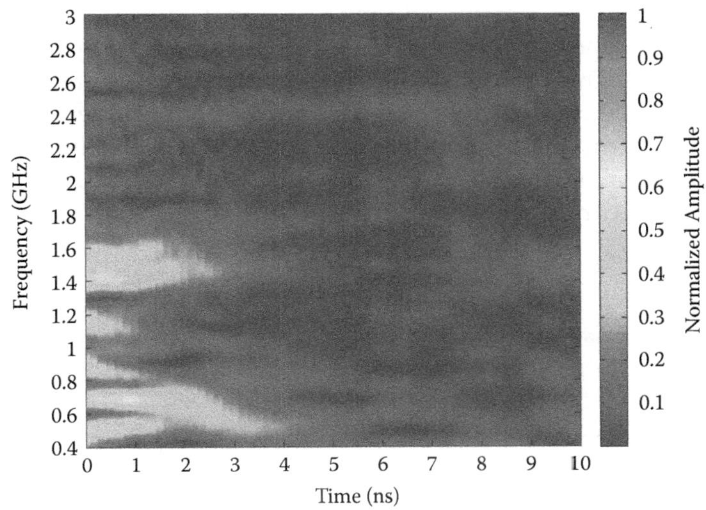

FIGURE 5.17
Time–frequency analysis for a human body without any concealed objects present. The amplitudes are normalized relative to the largest data point taken for the gun on body case (see Figure 5.18).

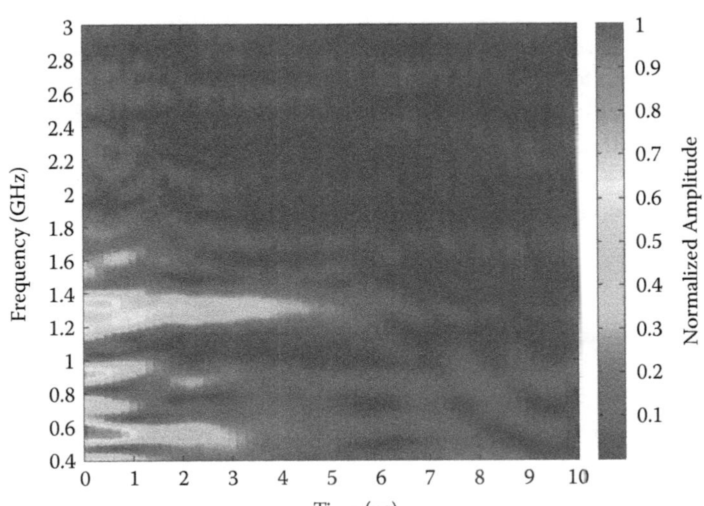

FIGURE 5.18
Time–frequency plot for a human body carrying a Glock handgun concealed on the abdomen. The contrast between the case for a body without a handgun is quite clear (see Figure 5.17).

certainly within the LTR of the body and handgun (Figure 5.18). Interestingly, the higher-frequency components of the LTR are enhanced by the presence of the handgun. This probably occurs because the presence of the handgun gives rise to an increased cross-polarized (vertically polarized) RCS. Thus, the EM waves travelling round the body are more effectively scattered in the vertical polarization state by the presence of the handgun, giving rise to enhanced higher-frequency components in the LTR.

5.4 Conclusion

The use of CNRs to detect the presence of objects, and particularly of weapons, of a size that can be easily concealed on the human body (<1 m) is feasible [46]–[50]. However, there are some considerable technical barriers to realizing a system that is efficacious enough for deployment as a screening device. An example of where such a system might be deployed is in an airport as a screening portal that might replace or complement current metal detector archways and active millimetre-wave imaging systems such as the L3 Provision. The inherent diffraction of microwaves of several hundred MHz around the human body prevents the LTR of a person being composed entirely of the LTR of any objects carried. In fact, the LTR from the human body is often considerably larger than that from a concealed handgun or knife and is unpredictable in frequency content, preventing the effective removal of the body's contribution. Further difficulties arise from the necessity of removing the scattered contribution from objects in the immediate vicinity of any person being screened by microwave excitation of CNRs. Reflections from background clutter (for example, other people walking past) can be time gated if they arise from objects that are separated from the person by a sufficient distance, so that the reflections do not overlap the LTR from the body.

References

1. C. V. Nelson, C. B. Cooperman, W. Schneider, D. S. Wenstrand, and D. G. Smith, Wide bandwidth time-domain electromagnetic sensor for metal target classification, *IEEE Transactions on Geoscience and Remote Sensing*, 39(6), 1129–1138, 2001.
2. C. V. Nelson, Wide-area metal detection system for crowd screening, presented at Proceedings of SPIE AeroSense 2003 Conference, Sensors and Command, Control, Communication, and Intelligence (C3T) Technologies for Homeland Defense and Law Enforcemnt II, Orlando, FL, 22–25 April 2003.

3. C. V. Nelson, Metal detection and classification technologies, *Johns Hopkins APL Technical Digest*, 24(1), 62–66, 2004.
4. N. G. Paulter, *Guide to the technologies of concealed weapon and contraband imaging and detection*, NIJ Guide 602-00, Electricity Division, National Institute of Standards and Technology, Gaithersburg, MD, February 2001. Prepared for National Institute of Justice Office of Science and Technology, Washington, DC.
5. A. Agurto, Y. Li, G. Yun Tian, N. Bowring, and S. Lockwood, A review of concealed weapon detection and research in perspective, presented at Proceedings of the 2007 IEEE International Conference on Networking, Sensing and Control, London, 15–17 April 2007.
6. A. Agurto Goya, New proposal for the detection of concealed weapons: electromagnetic weapon detection for open areas, PhD thesis, Huddersfield, UK, 2009.
7. A. Elgwel, S. W. Harmer, N. J. Bowring, and S. Yin, Resolution of multiple concealed threat objects using electromagnetic pulse induction, *Progress in Electromagnetics Research M*, 26, 55–68, 2012. doi:10.2528/PIERM12061113.
8. C. E. Baum, The singularity expansion method, in *Transient electromagnetic fields*, ed. L. B. Felson, 129–179. Spring Verlag, New York, 1976.
9. A. J. Berni, Target identification by natural resonant estimation, *IEEE Transactions on Aerospace and Electronic Systems*, AES-11(2), 147–154, 1975.
10. E. Rothwell, Extraction of the natural frequencies of a radar target from a measured response using E-pulse techniques, *IEEE Transactions on Antennas and Propagation*, 35(6), 715–720, 1987.
11. K. R. Menon, N. Balakrishnan, and K. Ramchand, Discriminating complex bodies with minor variations using e-pulse, *IEEE Transactions on Aerospace and Electronic Systems*, 44(1), 408–416, 2008.
12. R. Toribio, J. Saillard, and P. Pouliguen, Identification of radar targets in resonance zone: E-pulse techniques, *Progress in Electromagnetics Research (PIER)*, 43, 39–58, 2003.
13. L. O. Upton and L. A. Thurman, Radars for the detection and tracking of cruise missiles, *Lincoln Laboratory Journal*, 12(2), 355–366, 2002.
14. C. E. Baum, The singularity expansion method: background and developments, *IEEE Antennas and Propagation Society Newsletter*, August 1986.
15. C. E. Baum, The singularity expansion method and its application to target identification, *Proceedings of IEEE*, 79(10), 1991.
16. D. L. Clark, Early advances in radar technology for aircraft detection, *Lincoln Laboratory Journal*, 12(2), 2000.
17. A. Zyweck and R. F. Bogner, Radar target classification of commercial aircraft, *IEEE Transactions on Aerospace Electronic Systems*, 32(2), 598–606, 1996.
18. J. Chauveau, N. de Beaucoudrey, and J. Saillard, Characterization of perfectly conducting targets in resonance domain with their quality of resonance, *Progress in Electromagnetic Research (PIER)*, 74, 69–84, 2007.
19. M. Secmen and G. Turhan-Sayan, An electromagnetic target classification method for the target sets with alien target: application to small-scale aircraft targets, *Progress in Electromagnetics Research (PIER)*, 6(5), 425–429, 2010.
20. S. Harmer, N. Bowring, D. Andrews, and N. Rezgui, Ultra wide band aspect independent detection of on body concealed weapons using the out of plane polarised late time response, presented at 5th ESA Workshop on Millimetre Wave Technology and Applications and 31st ESA Antenna Workshop, 2009.

21. S. W. Harmer, S. E. Cole, N. J. Bowring, N. D. Rezgui, and D. Andrews, On body concealed weapon detection using a phased antenna array, *Progress in Electromagnetics Research (PIER)*, 124, 187–210, 2012.
22. S. W. Harmer, D. A. Andrews, N. D. Rezgui, and N. J. Bowring, Detection of handguns by their complex natural resonant frequencies, *IET Microwave Antennas Propagation*, 4(9), 1182–1190, 2010.
23. L. Zhang, Y. Hao, and C. G. Parini, Natural resonant frequency extraction for concealed weapon detection at millimetre wave frequencies, presented at 2nd European Conference on Antennas and Propagation (EuCAP 2007), Edinburgh, UK, November 2007.
24. A. Agurto and M. J. N. Sibley, New proposal for the detection of concealed weapons, presented at School of Computing and Engineering Researchers' Conference, University of Huddersfield, December 2007.
25. G. Mie, Beitrage zur optik truber medien, speziell kolloider metallosungen, *Annalen der Physik*, 4(25), 377–445, 1908.
26. Y. Wang, I. D. Longstaff, C. J. Leat, and N. V. Shuley, Complex natural resonances of conducting planar objects buried in a dielectric half-space, *IEEE Transactions on Geoscience Remote Sensing*, 39(6), 1183–1189, 2001.
27. M. Pasternak, J. Karczewski, D. Silko, W. Miluski, M. Lapinski, and P. Kaczmarek, Stepped frequency continuous wave GPR unit for unexploded ordnance and improvised explosive device detection, in *IEEE International Proceedings for Radar Symposium*, 2011, pp. 105–109.
28. S. Vitebiskiy and L. Carin, Resonance of perfectly conducting wires and bodies of revolution buried in a lossy dispersive half-space, *IEEE Transactions on Antennas and Propagation*, AP-44, 1575–1583, 1996.
29. Y. Wang, I. D. Longstaff, and C. J. Leat, Measurement of complex natural resonances of targets in free space and lossy media, *Progress in Electromagnetics Research (PIER)*, 29, 221–230, 2000.
30. Y. Hua and T. K. Sarkar, Generalised pencil-of-function method for the extracting poles of an EM system from its transient response, *IEEE Transactions on Antennas and Propagation*, 37(2), 1989.
31. J. H. Lee and H. T. Kim, Radar target discrimination using transient response reconstruction, *Journal of Electromagnetic Waves and Applications*, 655–669, 2005.
32. K. Kim, D. Seo, and H. Kim, Efficient radar target recognition using the MUSIC algorithm and invariant features, *IEEE Transactions on Antennas and Propagation*, 50(3), 325–337, 2002.
33. T. H. Shumpert, EMP interaction with a thin cylinder above a ground plane using the singularity expansion method, Air Force Weapons Laboratory, Sensor and Simulation Notes 182, June 1973.
34. J. S. Hey, G. S. Stewart, J. T. Pinson, and P. E. V. Prince, The scattering of electromagnetic waves by conducting spheres and discs, *Proceedings of the Physical Society B*, 69 1038, 1956.
35. L. A. Robinson, W. B. Weir, and L. Young, An RF time-domain reflectometer not in real time, *IEEE Transactions on Microwave Theory and Techniques*, 1(MTT-20), 855–857, 1972.
36. N. Geng, C. E. Baum, and L. Carin, On the low-frequency natural response of conducting and permeable targets, *IEEE Transactions of Geoscience and Remote Sensing*, 37(1), 1999.

37. C. E. Baum, Use of cross-pol to suppress early-time scattered signal, Interaction Notes, Note 609, 2 October 2008.
38. C. E. Baum, Polarimetric suppression of early-time scattering for late-time target identification, Note 599, University of New Mexico, December 2005.
39. O. Lopera, N. Milisavljevie, D. J. Daniels, and B. M. Macq, Time-frequency domain signature analysis of GPR data for landmine identification, in *IEEE Advanced Ground Penetrating Radar*, 2007, pp. 159–162.
40. E. Taoufik, S. Nabila, and B. Ridha, The reflection of electromagnetic field by body tissue in the UWB frequency range, in *IEEE International Radar Conference*, May 2010, pp. 1403–1407.
41. X. Li, E. J. Bond, B. D. Van Veen, and S. C. Hagness, An overview of ultra-wideband microwave imaging via space-time beam forming for early-stage breast-cancer detection, *IEEE Antennas and Propagation Magazine*, 47(1), 2005.
42. I. Craddock, M. Klemm, J. A. Leendertz, A. W. Preece, and R. Benjamin, Development and application of a UWB radar system for breast imaging, in *Antennas and Propagation Conference*, March 2008, pp. 24–27.
43. H. Lui and N. Shuley, Evolutions of partial and global resonances in transient electromagnetic scattering, *IEEE Antennas and Wireless Propagation Letter*, 7, 2008.
44. Y. Hua and T. K. Sarkar, Matrix pencil method for estimating parameters of exponentially damped sinusoids in noise, *IEEE Transactions on Acoustics, Speech, and Signal Processing*, 38, 814–824, 1990.
45. V. C. Chen and H. Ling, Joint time-frequency analysis for radar signal and image processing, *IEEE Signal Processing Magazine*, 16(3), 81–93, 1999.
46. A. Hunt and D. Hogg, Stepped-frequency CW radar for concealed weapon detection and through-the-wall surveillance, presented at SPIE Proceedings 4708, August 2002.
47. A. Vasalos, I. Vasalos, R. Heung-Gyoon, and S. E. Fotinea, LTR analysis and signal processing for concealed explosive detection, in *German Microwave Conference*, March 2010, pp. 166–169.
48. M. Gashinova, M. Cherniakov, and A. Vasalos, UWB signature analysis for detection of body-worn weapons, in *International Conference on Radar, CIE '06*, October 2006, pp. 1–4.
49. I. Vasalos, H. Ryu, and S. Fotinea, LTR analysis for fast concealed weapon detection, in *IEEE Microwave Conference (GeMIC)*, 2011, pp. 1–4.
50. M. Gashinova, M. Cherniakov, and L. Daniel, Late time response signature analysis in UWB sensors, in *Proceedings of ISAP2007*, Niigata, Japan, 2007, pp. 1242–1245.

6

Passive Millimetre-Wave Sensors

6.1 General Considerations

Operation of passive millimetre-wave sensors is based on an ability to detect the thermal radiation emitted by a human body. Since the main goal of the sensor is to detect any objects concealed on a human body, it is necessary to take into account absorption in clothing. It can be shown [1] that the total radiance emitted from a body screened by a fabric M_{rad} is

$$M_{rad} = \varepsilon_f \sigma \left(T^4_{body} - T_f^4\right) \tag{6.1}$$

where ε_f is the emissivity of fabric and $\sigma = 5.672 \times 10^{-8}$ Wm^{-2}K^{-4} is Stefan-Boltzmman's coefficient. The convection process in the gap between a human body and fabric is neglected. In order to estimate the net radiated power P_{net} we need to know the area of radiated surface A (about 2 m^2 for a human body). So,

$$P_{net} = A\, M_{rad} \tag{6.2}$$

Application of Wien's law to human body emission results in a peak wavelength of 9.5 μm [2]. The radiating power reduces with shift toward the millimetre (mm)-wave region. However, the sensitivity of modern mm-wave receivers is enough to capture the weak thermal radiation and use it for detection and imaging purposes. The efficiency of the detection depends also on the material composition of the object. Metallic objects need a temperature contrast between the human body (about 37°C) and background temperature (about 25°C) to be visible with a passive system, as they block all radiated emission from the body and completely reflect the thermal radiation of the surroundings. Nonmetallic objects such as plastics, explosives and ceramics may partially block the emission of the body, leading to reducing overall radiated power [3].

As an illustration, let us consider an idealized detecting scenario shown in Figure 6.1. The three cases possible in the scanning process are

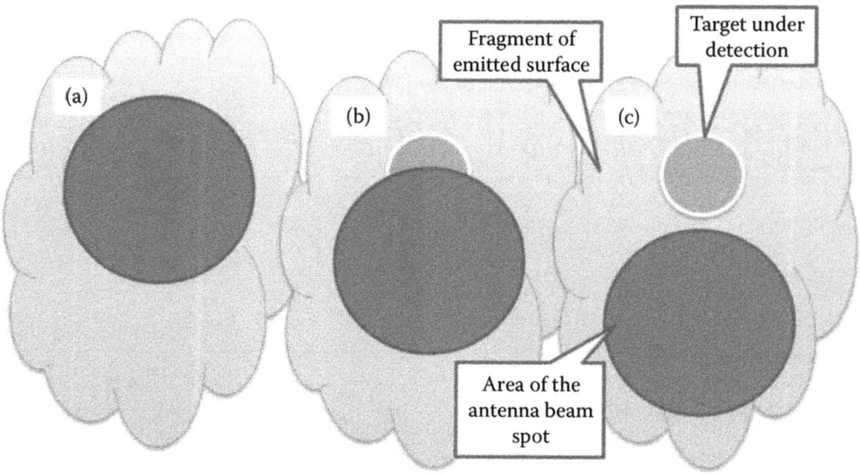

FIGURE 6.1
Three idealized detection scenarios.

- Antenna beam spot fully covers the detected target (Figure 6.1a).
- Antenna beam spot partially covers the detected target (Figure 6.1b).
- Antenna beam spot does not cover the target (Figure 6.1c).

It is assumed that the emitting surface has a constant intensity of radiated power and within the antenna's spot the intensity is uniformly distributed (i.e. the material is at a constant temperature and is uniformly composed of the same material). In the situation of Figure 6.1a the target partially blocks some of the power radiated. When the antenna's spot is shifted in position, as shown in Figure 6.1b, the target blocks less radiated power. In the position corresponding to Figure 6.1c, the antenna captures all the emitted power; none is blocked by the target.

The profile of the signal collected by antenna can be reproduced using the idealized geometrical structure shown in Figure 6.2, where a circular target of radius r is placed on a uniformly emitting surface scanned by antenna with the circular spot of radius R [9].

The points of intersections of both circles are the solution of the following system:

$$x^2 + y^2 = R^2$$

$$x^2 + (y - t)^2 = r^2 \tag{6.3}$$

where t is the shift of centre of the target along y-axis. The roots of Equation 6.3 can be written as follows:

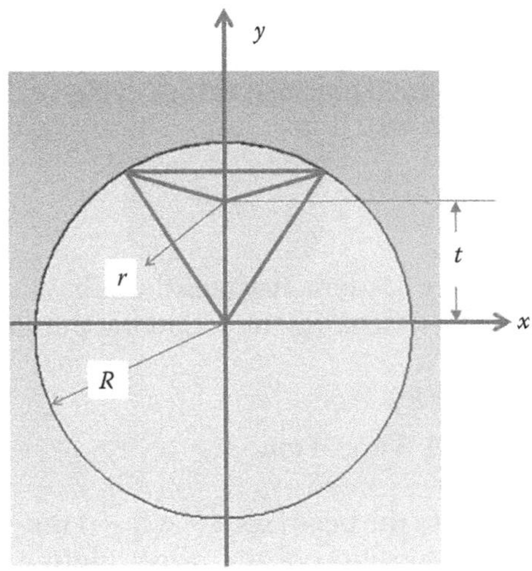

FIGURE 6.2
The idealized geometrical equivalent of the detection scenario shown in Figure 6.1. (From Kapilevich, B. et al., in *IEEE International Conference on Microwaves, Communications, Antennas, and Electronics Systems (COMCAS)*, Tel Aviv, Israel, 1–5, 2013. © 2013 by the Institute of Electrical and Electronics Engineers [IEEE].)

$$y = \frac{R^2 - r^2 + t^2}{2t}$$
$$x = \pm\sqrt{R^2 - (\frac{R^2 - r^2 + t^2}{2t})^2} \tag{6.4}$$

The blocking area of the target consists of two circular segments formed by the straight line connecting the intersection points and the circular arcs of radius R and r. The area of each segment is

$$AR = 0.5R^2(\theta_R - \sin\theta_R)$$
$$Ar = 0.5r^2(\theta_r - \sin\theta_r) \tag{6.5}$$

where θ_R and θ_r are the central angles corresponding to the arcs calculated from the following expressions:

$$\theta_r = 2\arcsin(x/R)$$
$$\theta r = 2(\pi - \arcsin(x/r)) \text{ if } R > |t| > r \tag{6.6}$$
$$\theta_r = 2\arcsin(x/r) \text{ if } |t| > R$$

As a result, the net blocking effect can be written as a sum of the aforementioned areas of segments and the normalized power density P_r^{norm} collected by the antenna shifted at the distance t, respectively, of the target's centre:

$$P_r^{norm} = \frac{[\pi R^2 - (A_R + A_r)]}{\pi R^2} \tag{6.7}$$

As an illustration, Figure 6.3 demonstrates the behaviour of P_r^{norm} as a function of the displacement t, assuming the following parameters:

- Radius of beam spot $R = 8$ cm
- Radius of target $r = 2$, 3 and 4 cm

When the target is inside the beam spot ($t < R - r$) the blocking effect does not depend on an actual position of the target (dotted lines). In the other extreme case, when the target is outside the beam spot ($t < R + r$), the antenna receives all the emitted power within its spot ($P_r^{norm} = 1$). In the positions of a target between these two extremes it partially blocks the emitted power. The difference between blocked and unblocked situations depends on the size of a target and decreases with decreasing target size. The ability of a passive sensor to distinguish this difference is determined by its spatial resolution.

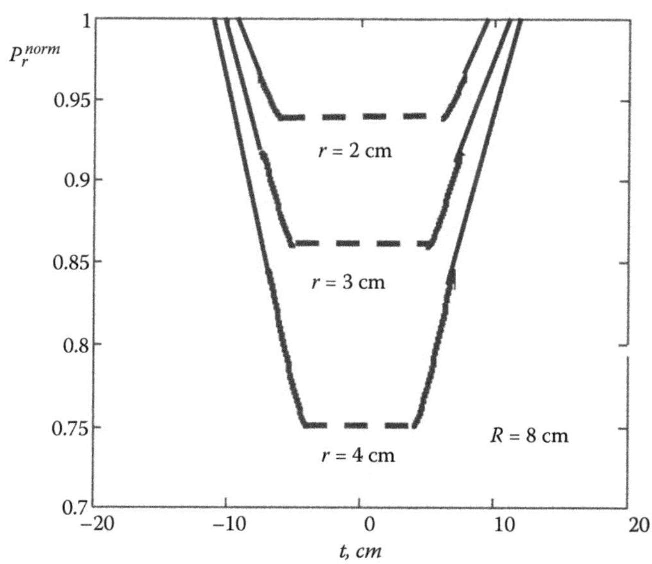

FIGURE 6.3
Idealized distribution of the normalized received power for various radii of blocking target. (From Kapilevich, B. et al., in *IEEE International Conference on Microwaves, Communications, Antennas, and Electronics Systems (COMCAS)*, Tel Aviv, Israel, 1–5, 2013. © 2013 by the Institute of Electrical and Electronics Engineers [IEEE].)

The latter is also a function of the diameter of a beam spot and the sensitivity of the receiver.

Based on the aforementioned considerations, we can formulate the general rule for detecting a concealed object using a passive mm-wave sensor. Namely, the uniformly emitting human body without a hidden object must demonstrate a unimodal distribution of the output signal in a process of single-traced scan from up-to-down (or left-to-right). The presence of the hidden object leads to violation of the unimodality condition in the output signal, producing a dip on the recorded trace. A practical realization of this approach has been implemented in the patent [4] and is discussed below.

6.2 Superheterodyne Receivers of Passive Sensors

The receiver of a passive sensor is a key component needed to capture energy emitted by a concealed object. By applying suitable signal processing techniques, it is possible to estimate the presence or absence of a threat object. Historically, such receivers had been designed at first for microwave and later for mm-wave radiometres. Ideally, they must have a wide bandwidth for operating with low-power quasi-noise signal emitted by a human body and high sensitivity since the emitted intensity is very low. For example, the average noise power per unit bandwidth emitted by a resistor at temperature 290 K is of the order of 10^{-20} W [5]. Since the typical microwave and mm-wave detectors need a minimum input of power of about –20 dBm, pre-amplification circuitry is essential to detect a low-level change in the input quasi-noise signals and to distinguish these variations. However, any amplification process necessarily injects additional noise, increasing the receiver noise temperature. Typically, the latter is higher than the equivalent temperature of the received signal and special modulation techniques are required for improving the overall system noise performance.

A classic example of the passive noise receiver is the Dicke radiometre [6]. A pulse-modulated signal is created by switching the receiver input between antenna and calibrated load, typically with a frequency of 50–1000 Hz. The output signal is proportional to the difference $T_A - T_c$, where T_A is the equivalent antenna temperature and T_c is the temperature of the calibrated load. Since the gain variations provide almost the same contribution to both measured components, the output signal is not dependent on such fluctuations, permitting one to reach an accuracy in measurement of temperature of the object of about 0.1 K or better.

For applications associated with detecting hidden objects, the knowledge of the absolute temperature of a blackbody is not critical, since the technique relies on the differential (contrast) temperature of the human body with and without contraband. So, the superheterodyne version of a Dicke

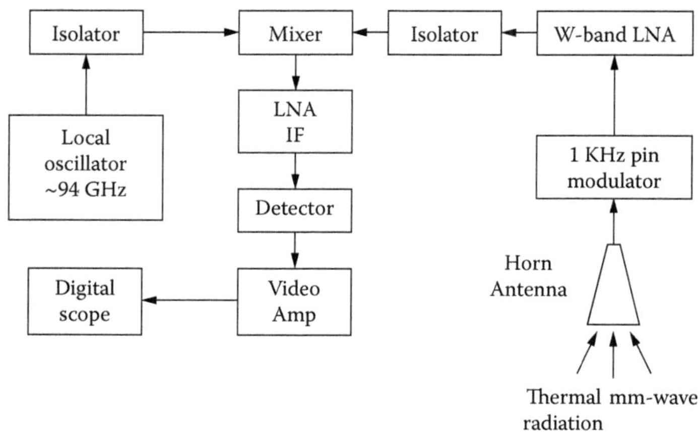

FIGURE 6.4
A modified superheterodyne version of a passive sensor.

noise receiver can be modified by removing the calibrated load as shown in Figure 6.4. Here, the receiving horn-lens antenna (gain 30 dBi and 3 dB band width 5°) is connected with the W-band pin diode modulator that provides 1 KHz pulse modulation of the received signal. Then the signal is amplified by a W-band low-noise amplifier (LNA) (gain about 20 dB) and down-converted by the mixer with an IMPATT local oscillator (LO) operating at 94 GHz. The intermediate frequency (IF) signal is amplified by about 70 dB by the two cascaded IF LNAs, with 6 GHz bandwidth, so that the total double-side bandwidth of the received signal is 12 GHz. A wide band HP-423A detector, post-detector video-amp, and Tektronix digital scope are employed for recording the output signal.

In the setup described here, a full W-band Farran noise source has been used for testing the receiver and estimating its overall noise performance. In order to compare the contribution of different circuit components to the overall system performance, both DC and AC output signals have been measured without averaging, as shown in Figure 6.5: V_{dc}, output DC voltage of the rectified noise; Δ, output AC voltage (peak to peak) of the same noise. In order to estimate the contribution of each element, they were independently switched off and on by means of the following step-by-step procedure:

- Video amplifier (see Figure 6.6a)
- Video amplifier + IF amplifiers (see Figure 6.6b)
- Video amplifier + IF amplifiers + local oscillator (see Figure 6.6c)
- Video amplifier + IF amplifiers + local oscillator + W-band LNA (see Figure 6.6d)
- Video amplifier + IF amplifiers + local oscillator + W-band LNA + Farran W-band noise source (see Figure 6.6e)

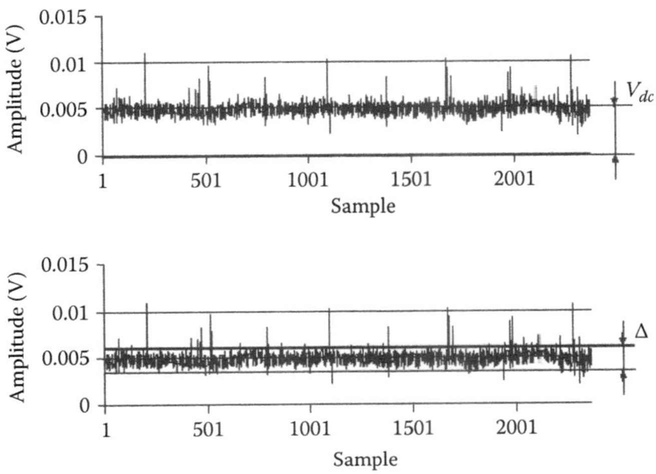

FIGURE 6.5
Illustration of the measured parameters: V_{dc}, output DC voltage of rectified noise; Δ, output AC voltage of the same noise.

Note that the last experiment has been carried out using the Farran W-band noise source with additional attenuation of –7.4 dB added in the output of the noise source to avoid the overload of the LNA.

A comparison of the measured data shown in Figure 6.6 clearly reveals that the IF amplifiers provide a major contribution to the total system noise performance. The IF circuitry consists of the two cascaded low-noise amplifiers needed to reach the total gain of 70 dB, resulting in increasing the total noise level of this part of the sensor.

Based on the above measurements we can estimate the noise performance of the designed mm-wave sensor using the Y-factor method [7]:

Estimation of the noise factor using DC output:

1. Output power ratio $Y = V_{dcout}/V_{dcin} = -6.56$ V/–4.72 V = 1.39
2. Actual ENR = 13 dB – 7.4 dB = 5.6 dB or E = 3.63
3. Noise factor $F = E/(Y - 1) = 3.63/(1.39 - 1) = 9.3$ or 9.7 dB

Estimation of the noise factor using AC output:

1. Output power ratio $Y = V_{dcout}/V_{dcin} = 480$ mV/360 mV = 1.33
2. Actual ENR = 13 dB – 7.4 dB = 5.6 dB or E = 3.63
3. Noise factor $F = E/(Y - 1) = 3.63/(1.33 - 1) = 11$ or 10.4 dB

Both approaches have demonstrated similar results. Figure 6.7 shows a general view of the passive sensor maintained on a programmable Az-El positioner.

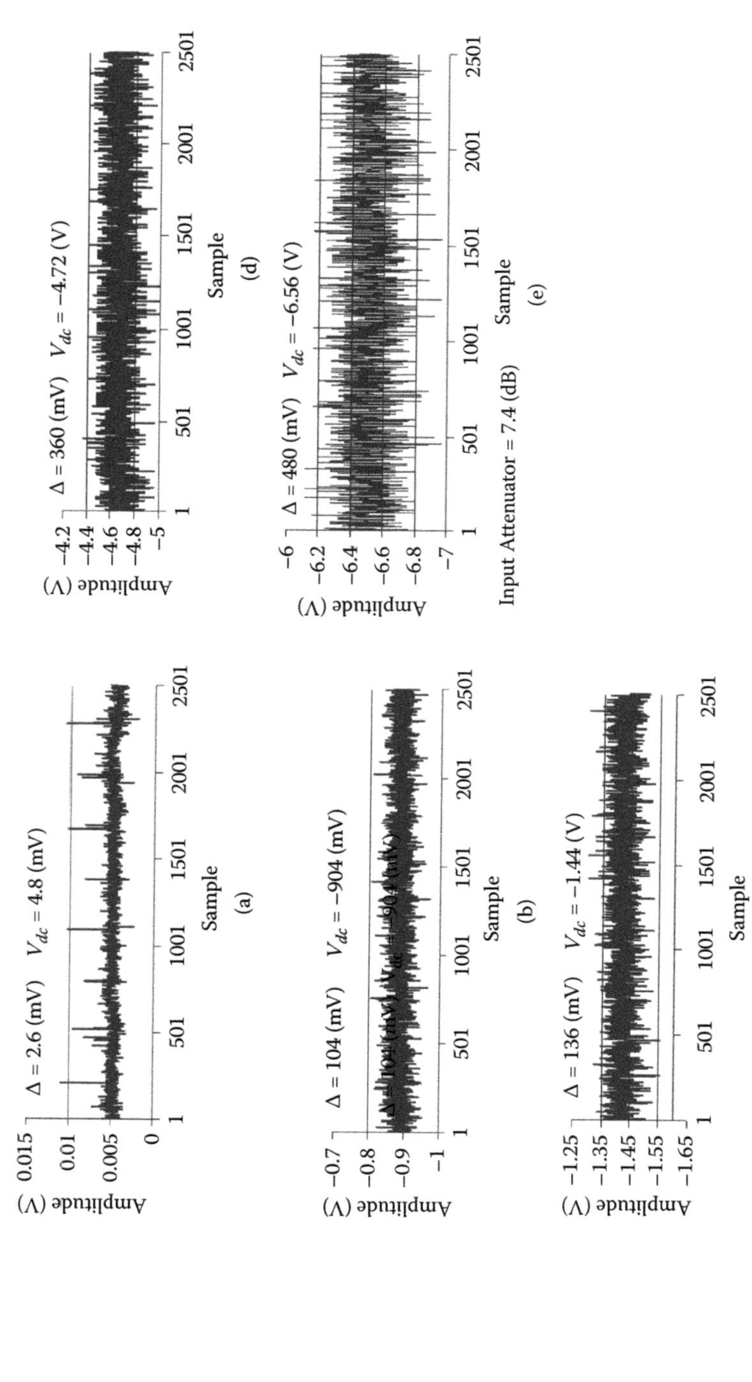

FIGURE 6.6
Illustration of the contribution of different elements of the mm-wave sensor to the system noise performance.

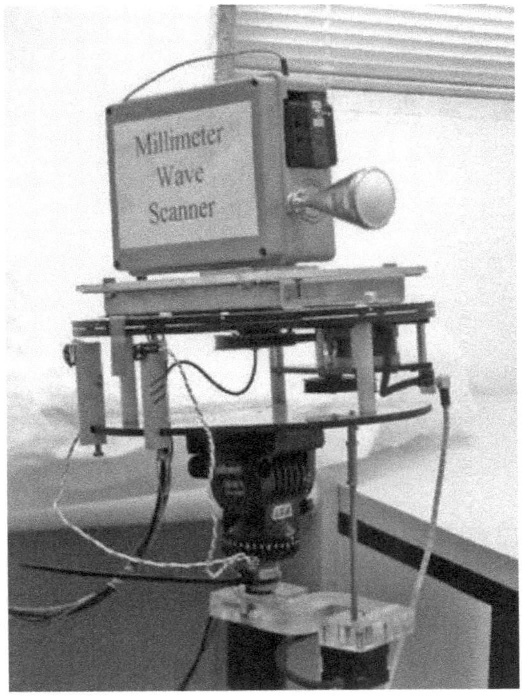

FIGURE 6.7
Photograph of the passive sensor maintained on a programmable Az-El positioner. (From Kapilevich, B. et al., in *Proceedings of the International Conference on Sensor Technologies and Applications*, Valencia, Spain, 20–23, 2007. © 2007 by the Institute of Electrical and Electronics Engineers [IEEE].)

Being combined with the positioner, the scanner is transformed to a single-pixel passive raster imager. An example illustrating its application for the indoor scanning scenario is depicted in Figure 6.8 [25]. A handgun on a human body (Figure 6.8a) has been scanned at *a* distance of about 80 cm away from the sensor and the corresponding passive mm-wave image is shown in Figure 6.8b. A noticeable thermal contrast between the open area of a body (the face or neck) and the area covered by a fabric can be clearly observed.

Figure 6.9 demonstrates the spatial resolution of the sensor for the indoor environment. Four metal balls of diameter 1.5 cm were placed at the vertex of a square rectangle (Figure 6.9a) having the side length 5 cm at a distance to the sensor of 75 cm. The passive mm-wave image of the balls using the sensor is shown in Figure 6.9b. The spatial resolution is approximately 1.5 cm/75 cm = 0.02 rad.

In order to extend the application of the sensor to outdoor conditions, the sensor is integrated with an antenna to give a higher gain, and therefore a smaller beam width. A good candidate is a Cassegrain parabolic dish [8] of

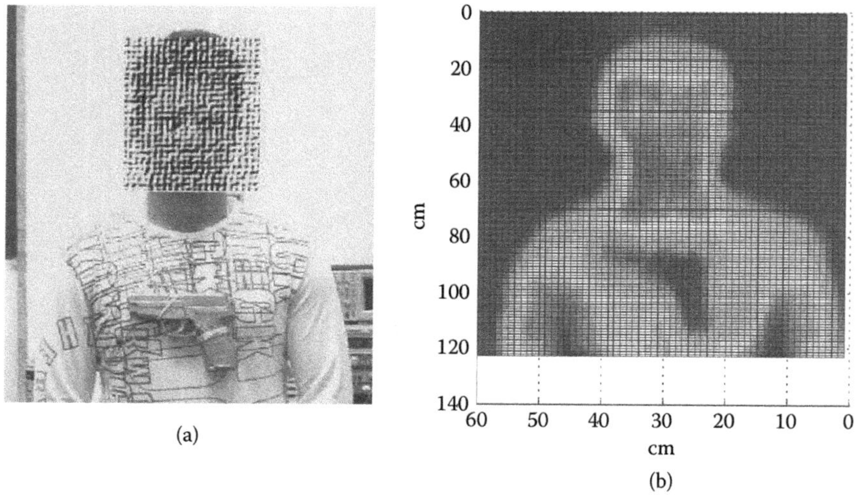

FIGURE 6.8
A handgun on a human body: (a) optical and (b) passive mm-wave images. (From Kapilevich, B. et al., in *Proceedings of the International Conference on Sensor Technologies and Applications*, Valencia, Spain, 20–23, 2007. © 2007 by the Institute of Electrical and Electronics Engineers [IEEE].)

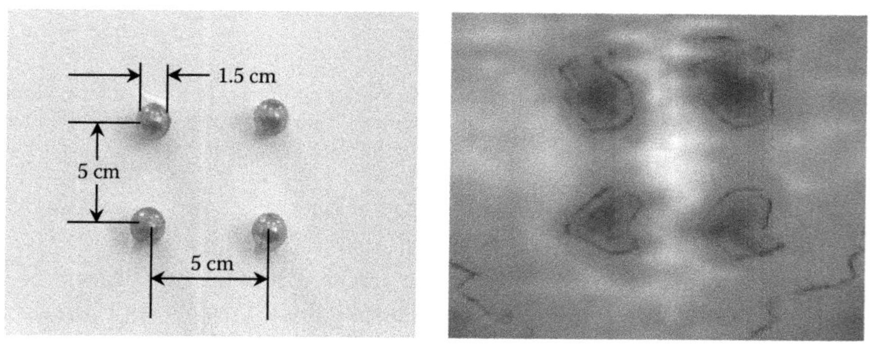

FIGURE 6.9
Four metal balls of diameter 5 cm: optical and passive mm-wave images.

diameter 30 cm with gain 45 dBi and a beam width of 0.7°. A view of a setup employing such an antenna is depicted in Figure 6.10.

Figure 6.11 demonstrates the 1D record of the signal at the output of the sensor shown in Figure 6.10 when the distance to cars is about 15 m. The peaks correspond to warmer areas corresponding to car engines. In order to illustrate an interaction of a human body with the warm object like a car's engine in outdoor conditions the scenery shown in Figure 6.12a has been used. Four different positions of the car and person, including the reference position without car and person, have been scanned azimuthally. The output signal of the sensor is different for all four positions, as shown in

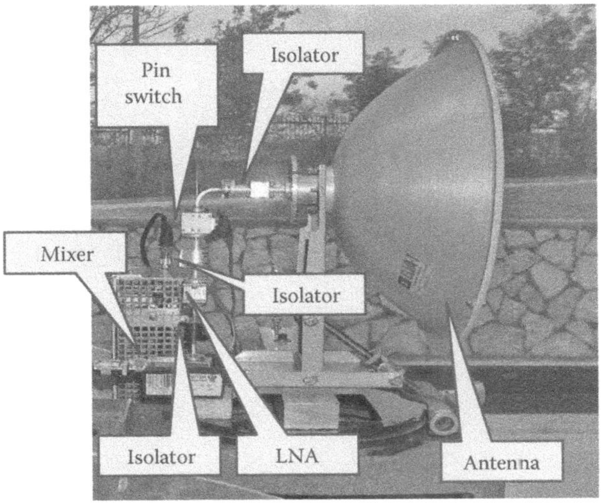

FIGURE 6.10
An outdoor version of a passive mm-wave sensor. (From Kapilevich, B. et al., in *IEEE 24th Convention of Electrical and Electronics Engineers in Israel*, 161–164, 2006. © 2006 by the Institute of Electrical and Electronics Engineers [IEEE].)

FIGURE 6.11 (See color insert.)
Output signals in the outdoor scenario with two cars present.

Figure 6.12b. It is clearly seen that in the presence of a person a portion of mm-wave energy emitted by the engine is blocked. The latter can be used for detecting and determining the position of the person passing the vehicle.

A further experiment using the same sensor has been carried out to demonstrate its operation at night. The mixed scenery with the streetlamp and

(a)

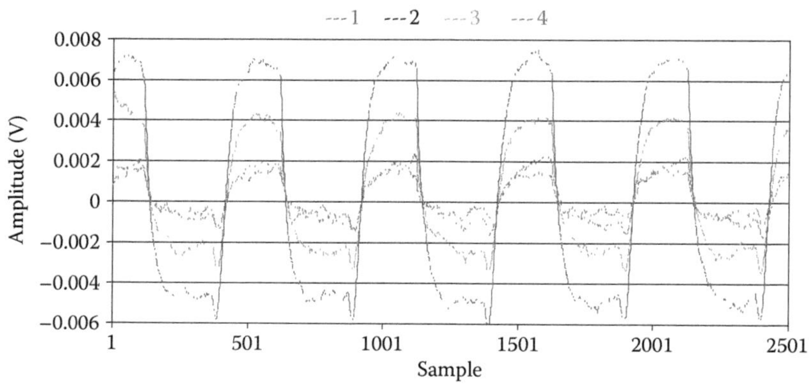

(b)

FIGURE 6.12 (See color insert.)
Mixed outdoor scenery: (a) a car and person and (b) recorded output signals. (From Kapilevich, B. et al., in *IEEE 24th Convention of Electrical and Electronics Engineers in Israel*, 161–164, 2006. © 2006 by the Institute of Electrical and Electronics Engineers [IEEE].)

bushes is shown in Figure 6.13a (optical daytime picture) and Figure 6.13b (optical nighttime picture) and after scanning at the night using the above passive sensor maintained on the programmable Az-El platform. The reconstructed mm-wave passive image is depicted in Figure 6.13c for a distance to the sensor of about 24 m. Assuming that the diameter of the top part of the streetlamp is approximately 12 cm, we can estimate the spatial resolution of the sensor, namely, 0.12 m/24 m = 0.05 rad.

6.3 Direct Detection Receivers as Passive Sensors

In order to reduce the overall cost of a passive mm-wave sensor, direct detection configuration of a receiver's module may be utilized. The general block diagram of such a module is shown in Figure 6.14. The emitted radiation collected by an antenna and modulated by a pin diode pulse switch is amplified by the wide band LNA.

A zero-bias GaAs Schottky diode detector is used for demodulation, and then the video signal goes to a processing unit. The output of the unit is

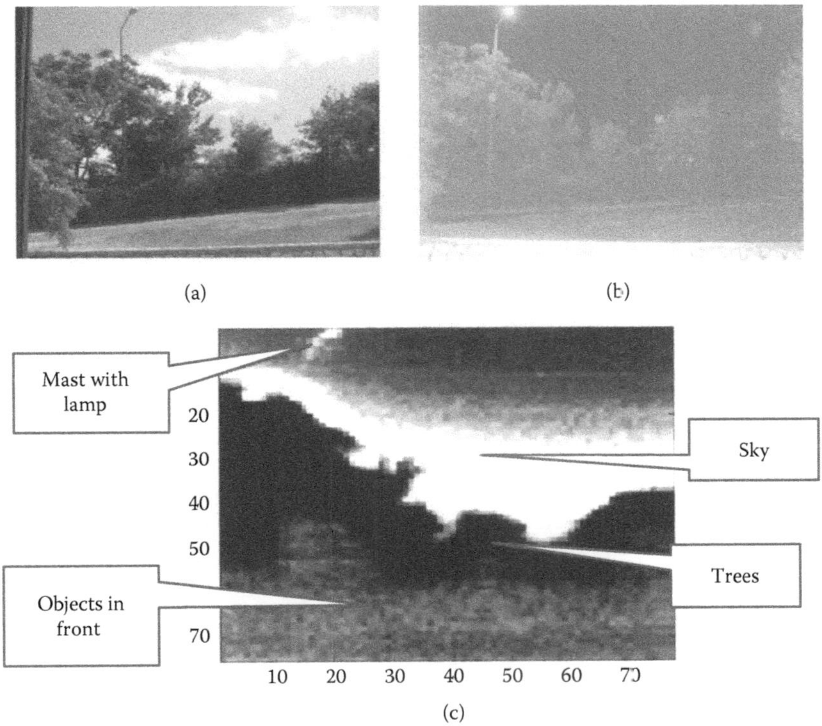

FIGURE 6.13 (See color insert.)
Mixed outdoor scenery: a streetlamp in the (a) day and (b) night and (c) the passive mm-wave image recorded at night.

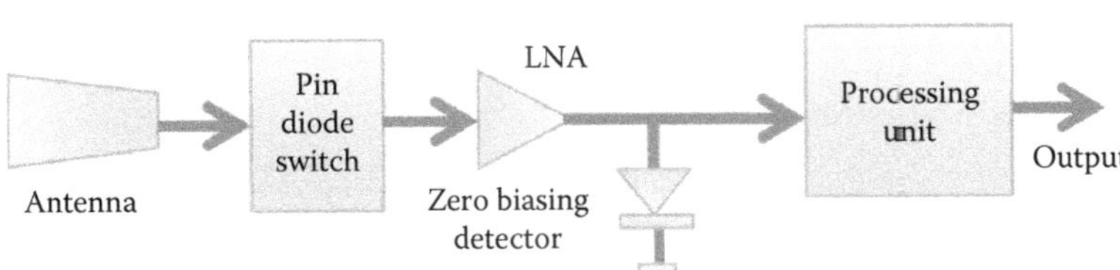

FIGURE 6.14
The block diagram of the direct detection receiver.

connected to an alarm indicator. The examples of direct detection receivers designed for mm-wave imagers were reported in [10], [11]. A similar configuration can be used for remote fire detection [12]. In order to achieve a noise equivalent temperature difference of 0.5 K or less, the gain of an LNA must be about 50 dB. Application of a more sensitive Sb-heterostructure tunnel diode permits the reduction of the gain of the LNA to 25–30 dB, which importantly minimizes the overall cost of the sensor discussed in [13].

The schematic of a passive mm-wave sensor designed for detecting obscured objects is shown in Figure 6.15 [26]. The signal emitted by a human

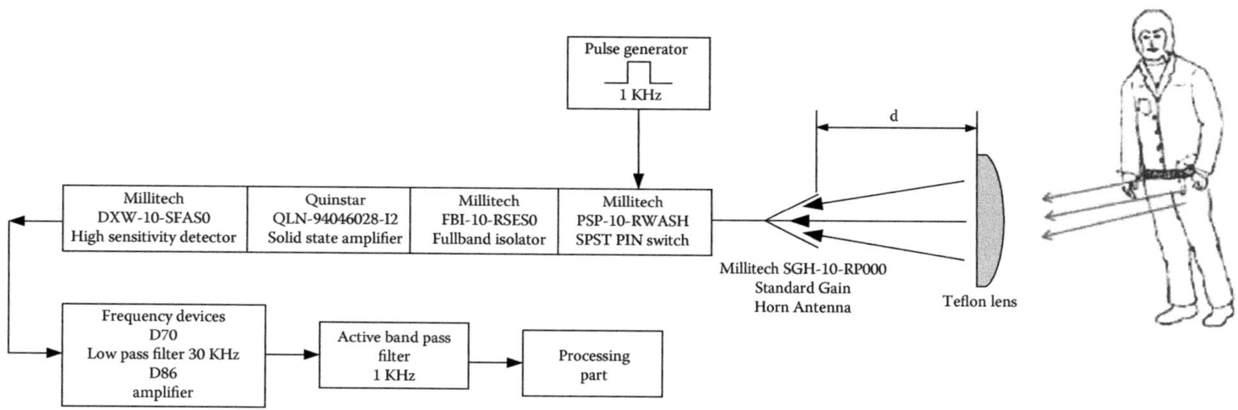

FIGURE 6.15
The schematic of the passive mm-wave sensor. (From Kapilevich, B. et al., *IEEE Sensors Journal*, 13, 11, 4224–4228, 2013. © 2013 by the Institute of Electrical and Electronics Engineers [IEEE].)

body with a hidden object is collected by the focusing Teflon lens placed at a distance *d* from the radiating aperture of a standard horn. This distance is varied to provide a focusing effect so the screened objects at different distances in the ranges 1.5–3 m can be detected with optimum spatial resolution. The signal is pulse modulated using a p-i-n switch driven by a pulse generator at the modulating frequency of 1 KHz. The full W-band ferrite isolator is added to provide matching with the LNA. A high-sensitivity zero-bias detector restores the video signal corresponding to the modulated frequency of 1 KHz. The latter passes through a low-pass filter integrated with the video amplifier. In order to improve the noise performance of the sensor, an active band pass filter, operating at the centre frequency 1 KHz, is added before the signal processing part.

Figure 6.16a shows the general photo of the passive sensor with three laser pointers around the perimeter of the focusing lens. The actual position of the antenna beam spot is inside the triangle formed by the laser point's vertexes (Figure 6.16b) where the scanning session is taking place in order to detect the hidden object, as shown in Figure 6.16c. Figure 6.17 depicts the output signals for different conditions:

1. The background level in the laboratory (Figure 6.17a)
2. The signal emitted by a person participating in the experiment at distance of about 2 m (Figure 6.17b)
3. The signal emitted by the same person with a hidden object, masking the normal returns from the body (Figure 6.17c)

The screenshot of the recorded signals is shown at the top left position of the figure. It is clearly seen that the absolute magnitude of the trace corresponding to a hidden object must be between the magnitudes corresponding to the background and 'open' human body. This feature can be used for developing a detection algorithm as described below.

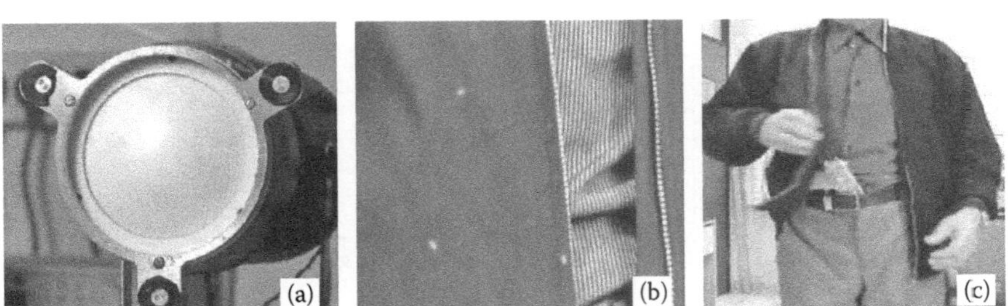

FIGURE 6.16 (See color insert.)
General picture of the sensor: (a) Teflon focusing lens and three laser pointers in front; (b) position of the beam spot during a scan indicated by laser pointers; (c) example of a concealed weapon. (From Kapilevich, B. et al., *IEEE Sensors Journal*, 13, 11, 4224–4228, 2013. © 2013 by the Institute of Electrical and Electronics Engineers [IEEE].)

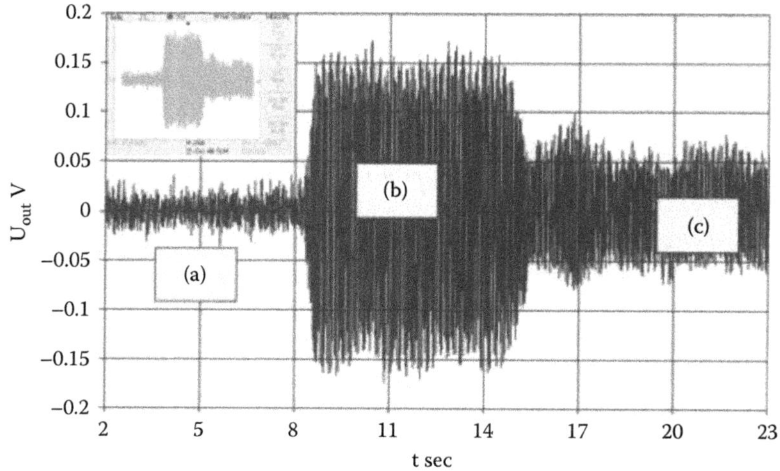

FIGURE 6.17
Example of the recorded output signals. (From Kapilevich, B. et al., *IEEE Sensors Journal*, 13, 11, 4224–4228, 2013. © 2013 by the Institute of Electrical and Electronics Engineers [IEEE].)

In order to estimate the operating range of the sensor, the output voltage as a function of distance has been measured when a human body emits thermal millimetre-wave radiation from the front and back surfaces. The results are almost identical, as is expected. The operating range depends on the sensor's noise level—about 0.02 V in the described sensor's case, which corresponds to an effective operational range of about 5 m (Figure 6.18). Assuming that the sensor is required to detect a target of minimum diameter 5 cm, an antenna

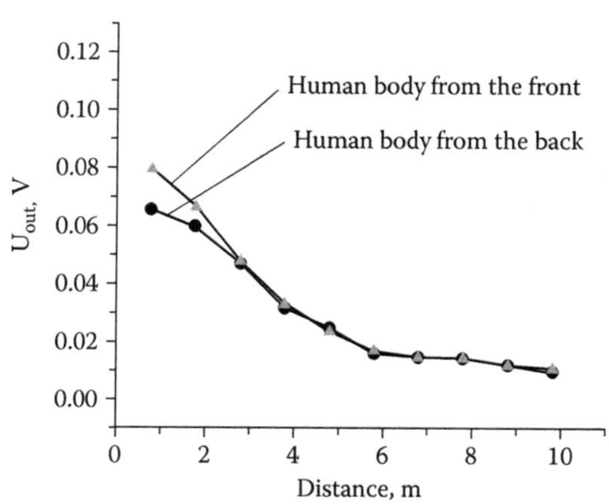

FIGURE 6.18
The measured output voltage as a function of distance when a human body emits thermal millimetre-wave radiation from the front and back surfaces. (From Kapilevich, B. et al., *IEEE Sensors Journal*, 13, 11, 4224–4228, 2013. © 2013 by the Institute of Electrical and Electronics Engineers [IEEE].)

with an angular resolution of about 0.6° should be employed. Such angular resolution needs a parabolic dish antenna with a diameter of about 30 cm, so that a practical realization of the portable sensor becomes quite problematic. As a compromise solution, a horn antenna with focusing lens was used allowing an operating range of up to 3 m with acceptable resolution.

Screening for concealed objects includes the following two phases:

1. Positioning the hot spot—the antenna must be directed at a point on the body which is not obscured (for example, the neck) in order to determine the maximum expected output signal.
2. Searching the obscured object position by moving the antenna using any specific trace preferred by the operator (typically manual; an up-down movement).

As a result of such a search the trace of output signal can be recorded and analyzed. Figure 6.19 shows examples of five independent traces (up-down) recorded at a distance 3 m from the following targets:

1. Human body
2. A dummy explosive semtex (10 × 15 cm^2) concealed on a human body under clothing
3. A handgun concealed on a human body under clothing

All traces were smoothed using a moving average algorithm. Within sample numbers 100–300, corresponding to the area of actual search of the human body background, the trace demonstrates unimodal behaviour, while the other two traces are non-unimodal due to the dips caused by the presence of the concealed objects.

At least three basic levels of millimetre-wave radiation intensity could be specified within the scanned field of view, as shown in Figure 6.20:

1. Background of thermal millimetre-wave radiation when the antenna is looking away from the region of interest, i.e. at a surrounding area containing clutter. The threshold level C3 must be set higher than this background intensity.
2. Radiation emitted by the human body when the antenna is looking directly at the open portion of a body (for example, the neck). The threshold C1 must be set a little bit lower than this intensity. As a result, the maximum dynamic range of the target detection is within the interval of the recorded output signal $C3 < V < C1$.
3. Radiation emitted by a human body and blocked by a hidden object when the antenna spot is looking directly at the actual position of the hidden object. The threshold C2 must be set higher than the lowest intensity and adjusted depending on the obscured object's properties.

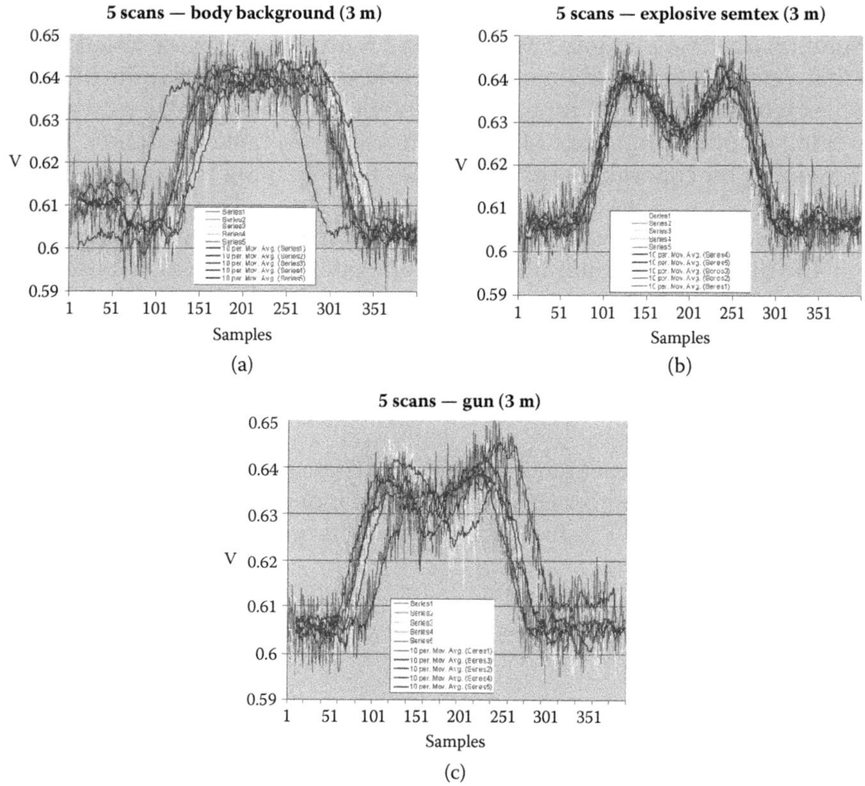

FIGURE 6.19
Sample recorded output traces: (a) the human body background, (b) the explosive semtex, (c) the handgun. (From Kapilevich, B. et al., in *IEEE International Conference on Microwaves, Commuications, Antennas and Electronic Systems (COMCAS)*, Tel Aviv, Israel, 1–5, 2013. © 2013 by the Institute of Electrical and Electronics Engineers [IEEE].)

The depth of the dip caused by the presence of the hidden object depends on its angular size relative to the size of the antenna spot, which in turn is determined by the spatial resolution of the antenna system. A horn antenna with focusing lens of diameter 9 cm was used, giving a 3 dB beam width of about 3°. So, the diameter of the beam spot at a distance of 3 m is about 15 cm. However, with the processing unit described below, the variation of the trace's magnitude can be registered more accurately, approximately 0.5 dB in our case. Therefore, the actual angular resolution is somewhat better, allowing detection of a target with a size of about 5 cm at a distance of 3 m.

Another factor limiting the sensor's resolution is the transparency of the detected target to millimetre-wave emission generated by the human body. However, most explosives and all metal samples have a sufficiently high attenuation resulting in sufficient thermal contrast for successful detection.

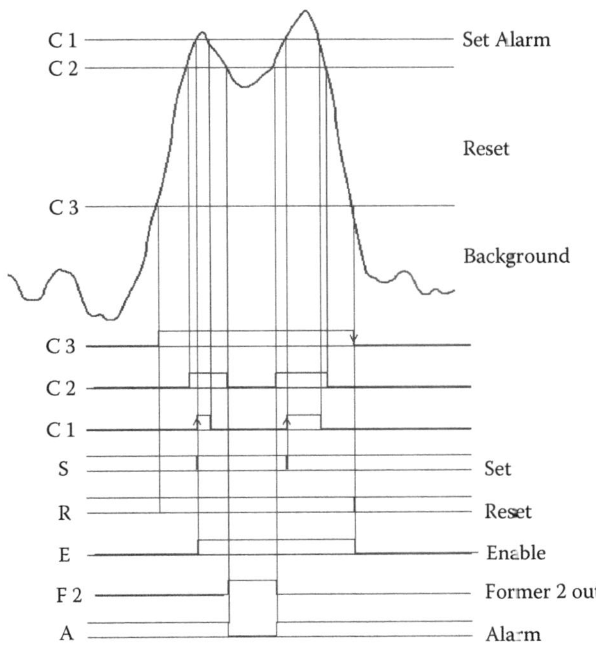

FIGURE 6.20
Typical traces in the presence of concealed objects recorded within the scanned field of view. (From Kapilevich, B. et al., *IEEE Sensors Journal*, 13, 11, 4224–4228, 2013. © 2013 by the Institute of Electrical and Electronics Engineers [IEEE].)

A block diagram of the processing unit is shown in Figure 6.21. The AC signal, proportional to the thermal intensity of millimetre-wave radiation received by the antenna, comes to the rectifier, and then splits to three comparators. The comparators C1 and C3 using formers 1 and 3 produce SET (S) and RESET (R) signals for the trigger, and the trigger sets or cancels the ENABLE (E) state to the gate. The comparator C2, using former 2, produces the ALARM (A) signals to the gate. The combination of ALARM and ENABLE produces the OUT alarm signal, indicating that an object is detected.

A signal coming to the processing part during a scanning session increases from the background level, and then crosses the C3 threshold level (Figure 6.20). Further movement of the antenna spot causes the signal to increase, which then consistently exceeds the threshold C2.

When the signal reaches the threshold C1, the comparator C1 switches from a low level (**L**) to a high level (**H**), and the former 1 produces a negative (**L**) pulse to switch the trigger to the SET position. The trigger sends an ENABLE signal to the gate. The gate now is ready to produce an alarm signal, while the former 2 changes its output from **L** to **H**.

This change happens when the signal crosses down below the threshold C2. The comparator C2 switches from **H** to **L**, the former 2 output signal

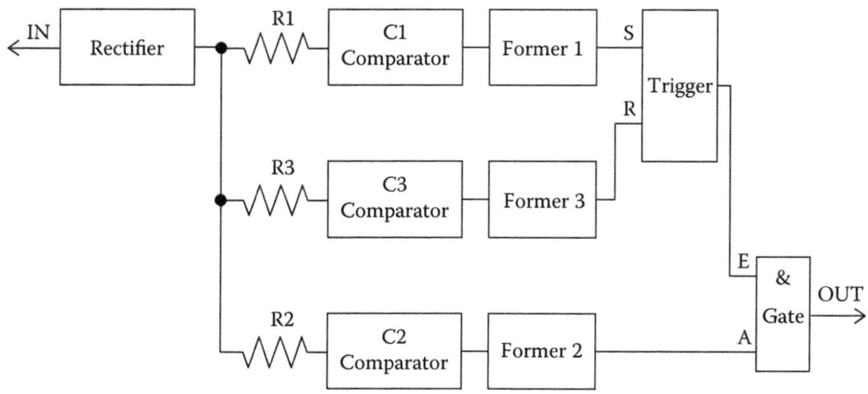

FIGURE 6.21
A block diagram of the processing unit. (From Kapilevich, B. et al., *IEEE Sensors Journal*, 13, 11, 4224–4228, 2013. © 2013 by the Institute of Electrical and Electronics Engineers [IEEE].)

switches from **L** to **H**, and the signal **A** becomes **H**; the gate will produce the output (OUT) alarm signal indicating the detection of a concealed object.

The ENABLE state of the gate remains until the signal drops back below threshold C3 (Figure 6.20). During the ENABLE time it is possible to make multiple scans of an obscured object(s) by manually moving the antenna beam within the area of the body to be inspected. In order to easily observe the current position of the antenna's spot, three point lasers are maintained around the perimeter of the focusing lens. Every time an obscured object is crossed by the antenna's beam the sound or light indicators are activated.

When the antenna spot is directed away from the body, and the signal drops below the threshold C3, the output of the comparator C3 will change to **L**, and therefore the system is reset automatically and ready for the next screening session. The details of the comparator's schematic have been described in [4].

The test experiments aimed at estimating the detection probability were performed with the metallic and dummy explosive samples listed in Table 6.1. All the tests were done in the lab (standard indoor environment), as shown in Figure 6.22. The distance between the person participating in the tests and the sensor was 3 m. The targets were placed on the human body under a jacket. Each target has been independently scanned by an operator (manually) five times and the averaged parameter $\delta = C_1 - C_2$ has been determined. The results of tests are summarized in Table 6.1 and demonstrate 100% detection probabilities. It should be pointed out that in real-world environments the detection probability will be lower due to undesirable interference caused by clutter. It is recommended to arrange an indoor checkpoint to be free of any such clutter effects in order to achieve optimal sensor performance.

TABLE 6.1

Summary of the Tests

Probability of False Alarm%	Probability of Detection (Relative Background Level)%	Averaged Parameter δ, V	Tested Target
0	100	0.025	CD 5 in.
0	100	0.01	Handgun
0	100	0.012	Pliers
0	100	0.01	C-4 (explosive)
0	100	0.015	DETA (explosive)
0	100	0.012	Semtex (explosive)

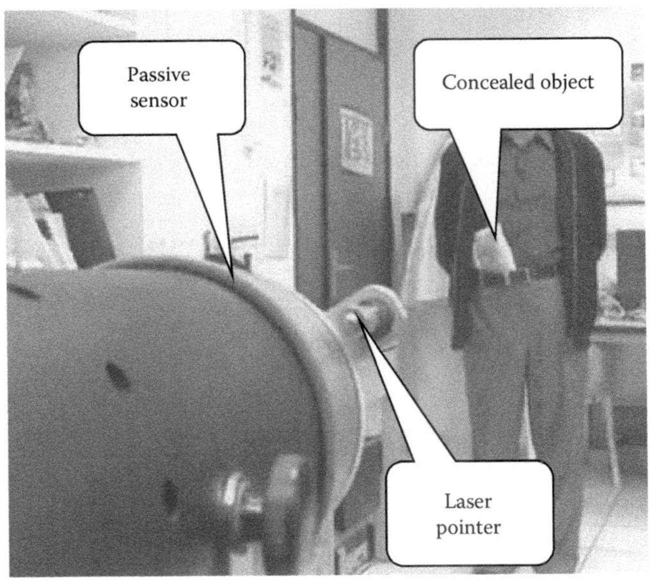

FIGURE 6.22
The scenery of the testing environment in the laboratory. (From Kapilevich, B. et al., *IEEE Sensors Journal*, 13, 11, 4224–4228, 2013. © 2013 by the Institute of Electrical and Electronics Engineers [IEEE].)

Independent laboratory tests of the sensor have been carried out with two objects hidden under clothing as shown in Figure 6.23 and the body without an object (empty body):

- Metal gun
- Dummy plastic explosive DETA
- Empty body

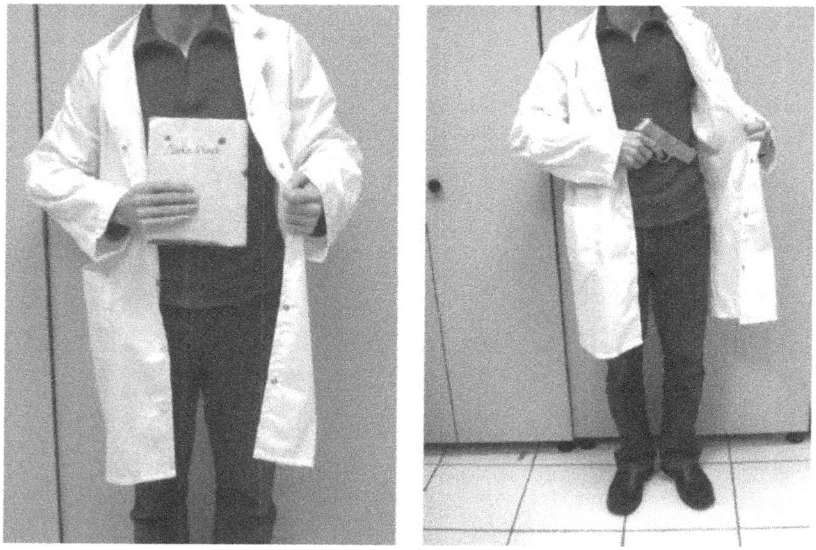

The indoor detecting hidden objects: "DETA" and gun.

FIGURE 6.23
Laboratory tests of the passive sensor.

TABLE 6.2
Summary of the Experiments Estimating the Probability of Detection

1	2	3	4	5	6	7	8	9	10	11	12	13	14	15	16	17	18	19	20
G	D	E	E	D	D	E	G	G	D	G	E	D	E	G	D	E	G	D	E
+	+	−	−	+	+	−	+	+	+	+	−	+	−	+	+	−	+	+	−
Detection probability 100%!																			

Note: G = hidden gun, D = hidden "DETA," E = empty body, (+) = hidden object is detected, (−) = hidden object is not detected.

Detection distance is about 2 m and number of trials is 20 for each tested object. The results of experiments are summarized in Table 6.2, demonstrating 100% detection in a laboratory environment. The prototyping version of the passive sensor can also be viewed here: https://www.youtube.com/watch?v=gIIL1VYUKV8.

6.4 Sensors Based on Correlation Receivers

One of the drawbacks of the described passive sensors is their sensitivity to environmental clutter. The deleterious effect of clutter is especially problematic in outdoor mode of operation. Sensors based on correlation receivers demonstrate better overall system performance and efficiency in detecting humans in outdoor conditions. Correlation receivers are widely used in radio astronomy [14] and satellite remote sensing [15]. They work by combining the outputs of individual receivers to form a synthetic aperture.

The simplified block diagram of the correlation receiver is shown in Figure 6.24. Each channel of the interferometre includes the antenna, input LNA and mixer driven by single LO. The output signal of each receiver consists of the two components: U_n (noise emitted by a black body) and U_s (signal from clutter), resulting in the total output signal $U_{total} = U_n + U_s$. Assuming that a sensor consists of two receiving channels, the multiplication of both outputs produces the following:

$$U_{total,1}U_{total,2} = (U_n + U_{s,1})(U_n + U_{s,2}) = U_{n},U_{s,2} + U_{s,1}U_n + U_{s,1}U_{s,1} + U_n{}^2 \quad (6.8)$$

Assuming that the noise components at the receiver's inputs are statistically independent, the first three terms in Equation 6.8 are eliminated after averaging, resulting in the only non-zero term,

$$\overline{U_{total,1}U_{total,2}} = \overline{U_n^2} \quad (6.9)$$

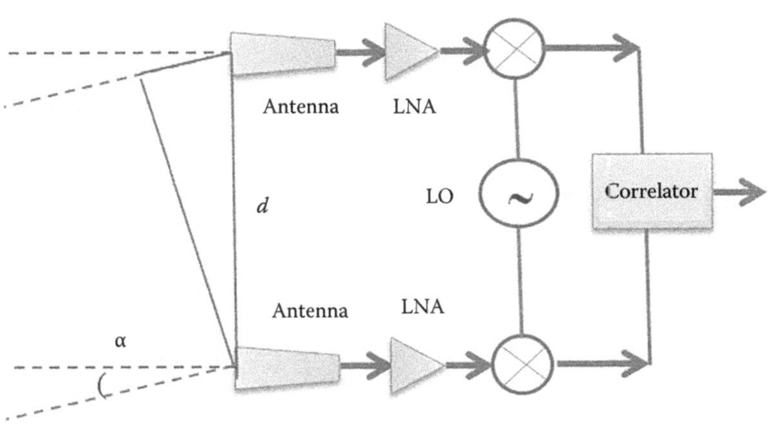

FIGURE 6.24
Simplified block diagram of the correlation receiver.

Equation 6.9 can be interpreted as the response of the ideal interferometre operating with infinite integration time. Since a human body radiates isotropically, the interferometre's output is maximized for the correlated signals and minimized for other sources, such as clutters with a lower level of correlation. The *I*/*Q*-components come to the inputs of the correlator, the output of which is proportional to the total received power.

The time delay of signals between the two antennae is equal to

$$\tau = \frac{d}{c} \sin\alpha \tag{6.10}$$

where c is the velocity of light. The interference pattern is a function of α, and if the baseline d is increased, the number of interference peaks (nulls) is also increased.

In order to estimate the received power in the most favourable scenario, the human body is assumed as an isotropic radiator. According to [16] a circular aperture antenna of 15 cm diameter permits a received power of about –91 dBm in the K_a-band from a human body at a distance of 15 m. Calculations were made for emissivity 0.64, surface area of skin facing the radiometre 0.9 m^2 and a receiver bandwidth 0.5 GHz.

Details of practical realization of passive interferometric sensors operating in the K_a-band were reported in [16]–[18]. The sensor can be also applied for detection of the angular velocity of a moving human, which is problematic for the sensor based on the Doppler effect [19].

Some related applications of quasi-passive sensors using noise-radiated signals in the X-band were also reported in [20], [21].

6.5 Sensors Based on Interference Effects

A second passive millimetre-wave technique might also be employed for security screening for low-loss dielectric layers, for example, very thin explosive sheets. The presence of dielectric materials between the body and the sensor (see Figure 6.25) will modulate the power received with a distinctive frequency dependence by virtue of constructive and destructive interference effects within the layers of dielectric material [22]. This effect can be measured by employing a sensor with a wide sensing bandwidth and dividing the power received by the sensor into narrower frequency bands or 'bins'. In this way, a human body, without appreciable clothing layers or other concealed materials, would thermally emit power at millimetre-wave frequencies proportional to the brightness temperature (product of absolute temperature and emissivity) of the body.

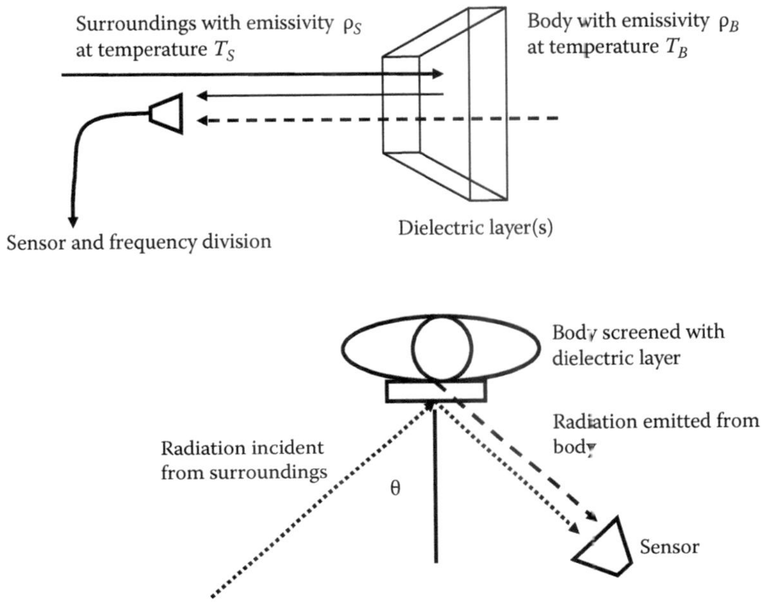

FIGURE 6.25
An illustration of the frequency modulation of passively emitted millimetre-wave radiation by passage through a low-loss dielectric slab of material.

The emissivity of the human body, ρ_b, has a frequency dependence [23], [24] which changes quite slowly with frequency in the millimetre-wave region of the spectrum. Additionally, the effects of the millimetre-wave radiation emitted from the surroundings and reflected from the human body and the directivity of the sensor are also required for a realistic model.

The sensor for the task of screening by measuring the thermal passive millimetre radiation emitted and reflected from a human body requires an aperture sufficient to provide a beam size on the body small enough that radiation from a body alone is sampled. If the beam spot is too large, then thermal radiation from the surroundings may also be measured, giving spurious results and rendering the screening operation unreliable. The aperture size, L, may be easily estimated by using the far-field angular diffraction limit which is given by Equation 2.12 and by writing the area of the sensor aperture as $A = L^2$.

If the sensor is required to operate out to a range R, then the minimum aperture size will be given by $L \approx Rc/w\nu$, where w is the maximum acceptable beam width of the sensor. The maximum acceptable beam width depends upon the minimum size of concealed object that is to be detected. A suitably high directivity antenna can be made from a Cassegrain reflector or a Gaussian optic lens.

The power at the sensor, in the absence of any screening layers, can be written as

$$P \approx \frac{\pi k \Delta\nu}{2}\left(\rho_S T_S R(\theta) + T_B\left(1 - R(\theta)\right)\right) \tag{6.11}$$

(see Equation 2.19), where the temperature of the surroundings is T_S, the human body temperature is T_B, the emissivity of the surroundings is ρ_S, the body has an angle-dependent reflectance, $R(\theta)$, the angle between the body normal and the axial direction of the detector is θ, and the centre frequency of the sensor is ν with a bandwidth of per frequency channel of $\Delta\nu$.

In the case of Equation 6.11 there are no dielectric layers between the sensor and the human body, and therefore the frequency dependence of the reflectance, $R(\theta)$, is not caused by interference effects within the screening layers, but instead is due to the changing complex permittivity of the human body with frequency. This dependence is slowly varying and therefore different in character to the effect caused by interference within layers. For example, the effect of interference within a single screening layer manifests as oscillation in received power which has a period in the frequency domain equal to $c/2\sqrt{\varepsilon}h$, where ε is the permittivity of the screening layer and h the thickness of that layer. Therefore, for the case of a single screening layer, an oscillatory pattern in the frequency domain with a fixed periodicity indicates the presence of a layer with an optical depth equal to $c/(2 \times \text{period})$. For multiple layers the signature is more complex but is still distinguishable from the case where there are no significant objects between the body and the sensor.

The introduction of planar layers of material between the sensor and the body serves to alter the frequency dependence of the power at the sensor receiver from the expression given by Equation 6.11 to that given by Equation 6.12:

$$P \approx \frac{\pi k \Delta\nu}{2}\left(\rho_S T_S R(\theta) + T_B t(\theta) + \sum_{m=1}^{M} A_m(\theta) T_m\right) \tag{6.12}$$

where the temperature of the individual screening layers is T_m. The screened body has an angle-dependent transmittance, $t(\theta)$, and each layer has angle-dependent absorptance, $A_m(\theta)$, which is equivalent to emissivity ρ_m (there are assumed $m = 1, 2, \ldots, M$ layers).

The frequency dependence of the quantities $R(\theta)$, $t(\theta)$ and $A_m(\theta)$ produces interference effects within the screening layers and will display a characteristic oscillatory pattern in the frequency domain, enabling the presence of layers in front of the body to be inferred.

Since conservation of energy requires that

$$R(\theta)+t(\theta)+\sum_{m=1}^{M} A_m(\theta)=1 \tag{6.13}$$

there will be no periodic frequency modulation if

$$T_m = T_B = \rho_S T_S \tag{6.14}$$

However, the condition given by Equation 6.14 is unlikely to be met in envisaged operational conditions where the human body, the surroundings and the screening layers are unlikely to share equal brightness temperatures. In the case of lossless dielectric screening layers, the ratio of the power received from a human body with and without screening materials takes on the simple form

$$\frac{P'}{P}=\frac{R'(\theta)+\dfrac{T_B}{\rho_S T_S - T_B}}{R(\theta)+\dfrac{T_B}{\rho_S T_S - T_B}} \tag{6.15}$$

The prime indicates the presence of the dielectric screening layers. $R'(\theta)$ may be smaller, equal or larger than $R(\theta)$ dependent on the thickness of the layers, the permittivity of the layers and the viewing angle. Equation 6.15 predicts that lossless, or nearly lossless, screening materials will not necessarily result in a reduction in the overall power received at the sensor antenna. In fact, under certain conditions the effective emissivity may be increased beyond that of a body alone.

The principle of operation for a frequency splitting antenna relies on thin-film interference effects within the low-loss dielectric screening layers [22]. Interference results in the incident, ambient mm-wave radiation (emitted from the surroundings) and the mm-wave radiation emitted from the body being modulated in the frequency domain after being reflected and transmitted, respectively, through the concealed layer(s). The form of frequency modulation depends upon the frequency dependence of the quantities $R(\theta)$, $t(\theta)$ and $A_m(\theta)$, which in turn depend upon the thickness, complex permittivity and angle at which the layers are viewed from by the sensor.

The reflectance, transmittance and absorptance of a planar stack of dielectric layers (see Figure 6.26) can be derived analytically as follows for both *p*- and *s*-polarized illumination. Note that in general the thermal grey body radiation emitted is unpolarized (that is, it consists of equal amplitudes of *p*- and *s*-polarized waves).

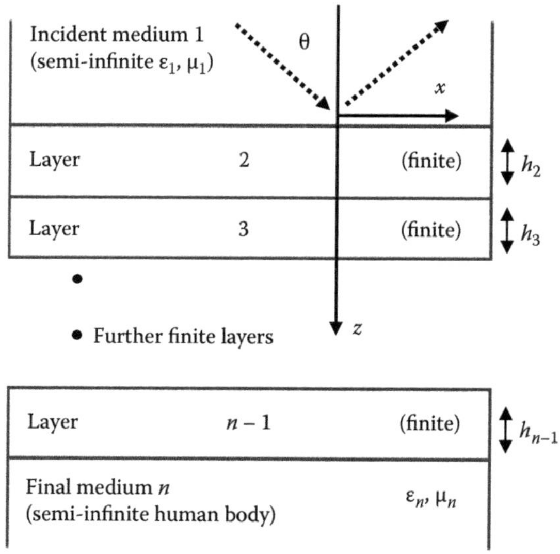

FIGURE 6.26
An illustration of the geometry of the model required for a general analysis of screening of the human body by a stack of n planar layers.

The electric field is arbitrarily chosen to exist in the y-direction only—that is, the s-polarized state. The tangential components of the electric and magnetic fields in the jth layer are

$$ {}_yE_j = \left({}_0E_{+j} \exp\left(i\kappa_j z\right) + {}_0E_{-j} \exp\left(-i\kappa_j z\right)\right) \exp i(k_x x - 2\pi\nu t) \tag{6.16} $$

$$ {}_xH_j = \frac{-\kappa_j}{2\pi\nu\mu_0\mu_j}\left({}_0E_{+j} \exp\left(i\kappa_j z\right) - {}_0E_{-j} \exp\left(-i\kappa_j z\right)\right) \exp i(k_x x - 2\pi\nu t) \tag{6.17} $$

where ${}_yE_j$ is the y-component of the complex electric field vector and ${}_xH_j$ is the x-component of the complex magnetic field vector. It is understood that the physical fields are represented by the real part of these expressions. ${}_0E_{+,j}$ and ${}_0E_{-,j}$ are the complex electric field amplitudes of the positive and negative travelling waves, with respect to the z-axis, in the jth layer. The complex z-component of the wave vector in the jth layer is given by κ_j, where

$$ \kappa_j = k_0\left(\varepsilon_j\mu_j - \varepsilon_1\mu_1 \sin^2\theta\right)^{1/2};\ k_0 = \frac{2\pi}{\lambda};\ c = \lambda\nu \text{ and } k_x = k_0\sqrt{\varepsilon_1\mu_1}\sin\theta $$

Here we use ν to denote the frequency of the radiation in cycles per second, λ to denote the wavelength of the radiation in free space, and c is the speed of light in free space.

By imposing suitable boundary conditions on the electromagnetic fields given by Equations 6.16 and 6.17, that of continuation of tangential components of the electric and magnetic fields across every interface, we have in the jth layer,

$$ {}_0E_{+,j}\exp(i\kappa_j h_j) + {}_0E_{-,j}\exp(-i\kappa_j h_j) = {}_0E_{+,j+1} + {}_0E_{-,j+1} \tag{6.18} $$

for the tangential electric field, and

$$ \frac{\kappa_j}{\mu_j}\left({}_0E_{+,j}\exp(i\kappa_j h_j) - {}_0E_{-,j}\exp(-i\kappa_j h_j)\right) = \frac{\kappa_{j+1}}{\mu_{j+1}}\left({}_0E_{+,j+1} - {}_0E_{-,j+1}\right) \tag{6.19} $$

for the tangential magnetic field. Using Equations 6.18 and 6.19 the complex electric field amplitudes for the positive and negative travelling waves in the j + 1th layer can be written as a matrix equation in terms of the analogous quantities in the jth layer.

$$ \begin{bmatrix} {}_0E_{+,j+1} \\ {}_0E_{-,j+1} \end{bmatrix} = \begin{bmatrix} (1+\chi_j)\exp(i\kappa_j h_j) & (1-\chi_j)\exp(-i\kappa_j h_j) \\ (1-\chi_j)\exp(i\kappa_j h_j) & (1+\chi_j)\exp(-i\kappa_j h_j) \end{bmatrix} \begin{bmatrix} {}_0E_{+,j} \\ {}_0E_{-,j} \end{bmatrix} \tag{6.20} $$

where

$$ \chi_j = \frac{\kappa_j \mu_{j+1}}{\kappa_{j+1}\mu_j} $$

is the ratio of the impedances in the j + 1th layer and the jth layer.

Equation 6.20 is a recurrence relation which may be used to determine the complex electric field amplitudes in the jth layer in terms of the incident electric field amplitudes in medium 1. From Equation 6.20 we obtain

$$ F_j = M_{j-1}M_{j-2}\ldots\ldots\ldots\ldots M_1F_1 \tag{6.21} $$

where

$$ F_j = \begin{bmatrix} {}_0E_{+j} \\ {}_0E_{-j} \end{bmatrix} $$

and

$$ M_j = \begin{bmatrix} (1+\chi_j)\exp(i\kappa_j h_j) & (1-\chi_j)\exp(-i\kappa_j h_j) \\ (1-\chi_j)\exp(i\kappa_j h_j) & (1+\chi_j)\exp(-i\kappa_j h_j) \end{bmatrix} $$

Since the product $M_{j-1}M_{j-2}..........M_1$ in Equation 6.21 is itself a 2 × 2 matrix, we write for brevity,

$$N_j = M_{j-1}M_{j-2}..........M_1$$

where we define

$$N_j = \begin{bmatrix} A_j & B_j \\ C_j & D_j \end{bmatrix}$$

Equation 6.21 can now be written succinctly as

$$F_j = N_j F_1 \tag{6.22}$$

In the final semi-infinite medium, labelled by the index n in Figure 6.26, there is no reflected wave, as there is no discontinuity to partially reflect the transmitted wave; therefore, from Equation 6.22 we can write

$$\begin{bmatrix} {}_0E_{+n} \\ 0 \end{bmatrix} = \begin{bmatrix} A_n & B_n \\ C_n & D_n \end{bmatrix} \begin{bmatrix} {}_0E_{+1} \\ {}_0E_{-1} \end{bmatrix} \tag{6.23}$$

Using Equation 6.23 we can write the complex reflectivity for the electric fields from the layered structure. For the reflectivity we have

$$\Gamma_n(\nu) = \frac{{}_0E_{-1}}{{}_0E_{+1}} = -\frac{C_n}{D_n} \tag{6.24}$$

and for the transmission coefficient,

$$T_n = \frac{{}_0E_{+n}}{{}_0E_{+1}} = A_n - \frac{B_n C_n}{D_n} \tag{6.25}$$

Using Equations 6.16, 6.17 and 6.22 gives expressions for the EM field distribution in the jth layer in terms of the incident electric field amplitude, ${}_0E_{+1}$:

$${}_yE_j = {}_0E_{+1}\left(\left(A_j + B_j\Gamma_n\right)\exp\left(i\kappa_j z\right) + \left(C_j + D_j\Gamma_n\right)\exp\left(-i\kappa_j z\right)\right)\exp i\left(k_x x - \omega t\right) \tag{6.26}$$

$${}_xH_j = \frac{-\kappa_j\,{}_0E_{+1}}{\mu_0\mu_j\omega}\left(\left(A_j + B_j\Gamma_n\right)\exp\left(i\kappa_j z\right) - \left(C_j + D_j\Gamma_n\right)\exp\left(-i\kappa_j z\right)\right)\exp i\left(k_x x - \omega t\right) \tag{6.27}$$

The irradiance I_j of the EM fields in the jth layer is given by the z-component of the Poynting vector, averaged over a time cycle

$$I_j = -\frac{1}{2}\mathrm{Re}\left\{{}_yE_{j,\,x}H_j^*\right\} \tag{6.28}$$

First, we wish to find out the reflectance, transmittance and absorptance of the structure; these quantities are determined by the appropriate ratios of the irradiances in the two semi-infinite media, labelled 1 and n. The z-component of the time-averaged Poynting vector for the incident EM wave is easily calculated using only the forward and backward travelling wave terms from Equations 6.16 and 6.17 and applying Equation 6.28.

$$I_{\pm 1} = \sqrt{\frac{\varepsilon_1}{\mu_1}}\frac{\left|{}_0E_{\pm 1}\right|^2}{2\mu_0 c}\cos(\theta) \tag{6.29}$$

The z-component of the time-averaged Poynting vector in the final, semi-infinite medium is similarly evaluated,

$$I_{+n} = \frac{\left|{}_0E_{+n}\right|^2}{2\mu_0 c\left|\mu_n\right|^2}\left(\mathrm{Re}\left\{\left(\varepsilon_n\mu_n - \varepsilon_1\mu_1\sin^2\theta\right)^{1/2}\right\}\mu_n' + \mathrm{Im}\left\{\left(\varepsilon_n\mu_n - \varepsilon_1\mu_1\sin^2\theta\right)^{1/2}\right\}\mu_n''\right) \tag{6.30}$$

The reflectance, transmittance and absorptance of the planar structure can be calculated from Equations 6.29 and 6.30. The reflectance is given by

$$R = \frac{I_{-1}}{I_{+1}} = \left|\frac{{}_0E_{-1}}{{}_0E_{+1}}\right|^2 = \left|\Gamma_n\right|^2 = \left|\frac{C_n}{D_n}\right|^2 \tag{6.31}$$

The transmittance of the structure is defined as $T = \frac{I_{+n}}{I_{+1}}$; therefore, the transmittance is

$$\mathrm{T} = \sqrt{\frac{\mu_1}{\varepsilon_1}}\frac{1}{\left|\mu_n\right|^2\cos\theta}\left|A_n - \frac{B_nC_n}{D_n}\right|^2 \left(\mathrm{Re}\left\{\left(\varepsilon_n\mu_n - \varepsilon_1\mu_1\sin^2\theta\right)^{1/2}\right\}\mu_n' + \mathrm{Im}\left\{\left(\varepsilon_n\mu_n - \varepsilon_1\mu_1\sin^2\theta\right)^{1/2}\right\}\mu_n''\right) \tag{6.32}$$

Finally the absorptance of the *whole* structure is given simply by

$$A = 1-(R+T) \tag{6.33}$$

and by applying Equations 6.26–6.29 we can find the absorptance of the jth layer, noting that the Poynting vector obtained will be a function of the distance z through the jth layer. The absorptance of the jth layer is simply given by

$$A_j = \frac{I_j(0) - I_j(h_j)}{I_{+1}} \tag{6.34}$$

The preceding analysis is valid only for s-polarized plane waves; the equations for p-polarized plane waves can be obtained directly from the equations for the s-polarized analysis with the following transformations applied:

$$E \to H, H \to -E \text{ and } \varepsilon_j\varepsilon_0 \leftrightarrow \mu_j\mu_0$$

Applying these transformations to the key results in the analysis gives the analogous equations for the p-polarized case. Under transformation the ratio of impedances given by χ_j is changed for p-polarization to

$$\chi_j = \frac{\kappa_j \varepsilon_{j+1}}{\kappa_{j+1} \varepsilon_j};$$

this change in turn alters the elements A_j, B_j, C_j and D_j that form matrix N_j.

6.6 Modelling Results

Using MATLAB®, a model which implements the previous mathematical descriptions is used to predict the magnitude of the interference effect produced by layers located in front of the human body. It is apparent that the background brightness temperature is a key factor in determining the magnitude of the predicted brightness temperature variations from the body. In Figure 6.27 three conditions, outdoors with a clear sky, outdoors with an overcast sky and indoors, are shown for a lossless layer placed in front of the body. The largest variation is presented with the low reflected radiation levels from the clear sky; in this case the dominant phenomenon is transmission of millimetre waves, emitted from the body, through the lossless layer, with modulation caused by interference effects for transmission. In the case of an overcast sky, where the incident millimetre-wave radiation levels are higher, but still less than those emitted from the body, the variation is somewhat reduced. The reduction comes from the fact that when transmittance of the layer is at a maximum, the reflectance must be at a minimum (as there is no loss within the layer) and vice versa. This cancelling

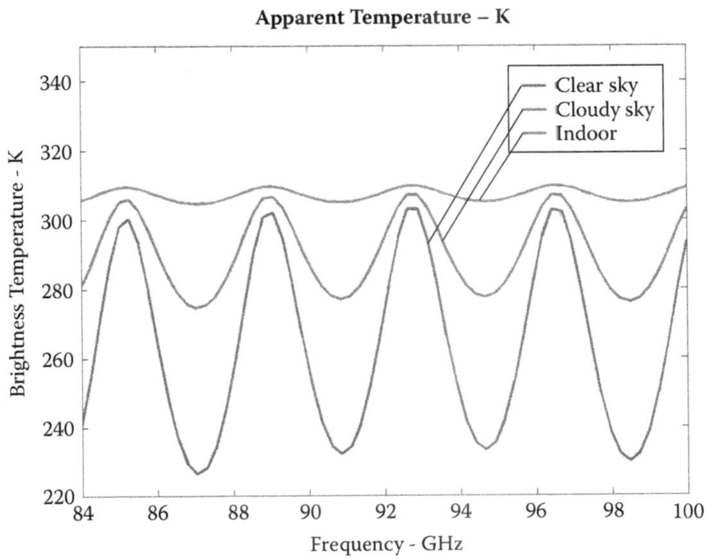

FIGURE 6.27
Predicted brightness temperature variation with frequency for three different external temperatures: 50, 200 and 293.15 K; corresponding to outdoor operation with a clear sky, overcast sky and indoor use at room temperature. The sensor is aligned at normal incidence to the person, and the person is carrying a 25 mm thick, lossless, dielectric layer positioned in front of the body. The layer has permittivity 2.5 and the temperature of the body is assumed to be 310.15 K.

effect becomes dominant in the indoor scenario, where the power emitted from the body is not much greater than the power incident from the surroundings. In this case, the effects do not cancel completely as there is still a difference in brightness temperature between the body and the surroundings (see Equation 6.14). The inclusion of loss in the layer complicates the matter, as now the maxima/minima of reflectance do not necessarily correspond directly with the minima/maxima of transmittance and the layer has an emittance, and therefore a brightness temperature, of its own. This is because power may be dissipated within the layer and not be reflected or transmitted. For quite low-loss layers the effect merely reduces the overall variation in temperature seen over the frequency range under all conditions (see Figure 6.28). In higher loss cases (see Figure 6.29), the effect is drastic and indoor operation becomes impossible, as the variation in temperature becomes too small to be a reliable indication; outdoors the effect may still be measurable. The blocking effect of the high-loss layers on the body emission is apparent when contrasting the lossless and high-loss cases (Figures 6.27 and 6.29), the mean brightness temperature values for the lossless case being significantly higher than the mean brightness temperature values for the high-loss case. This effect results from radiation being absorbed in the layer from both incident surroundings and the body, giving a lower brightness

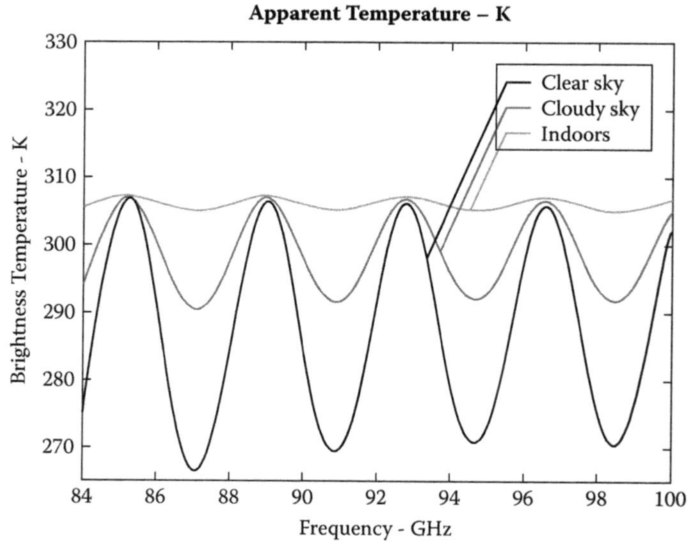

FIGURE 6.28
Variation of predicted brightness temperatures when the dielectric layer has moderate loss: complex permittivity of the layer is $2.5 + i2.5 \times 10^{-2}$. The layer has a temperature of 305.15 K, and the temperature of the body is 310.15 K.

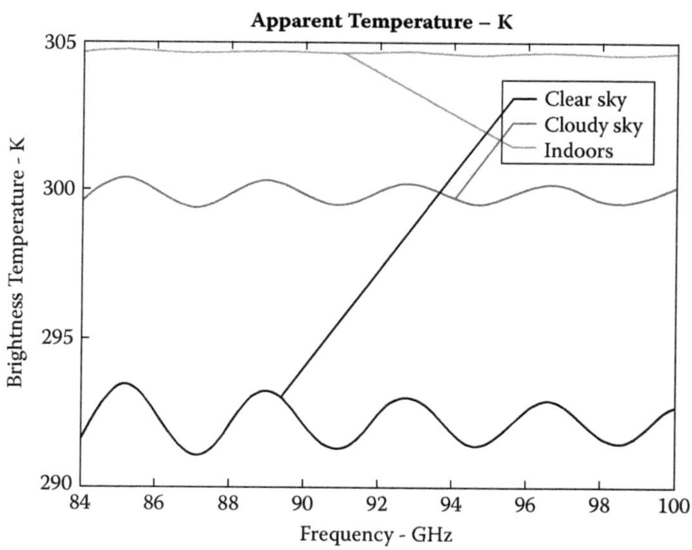

FIGURE 6.29
Variation of predicted brightness temperatures when the dielectric layer has high loss: complex permittivity of the layer is $2.5 + i1.25 \times 10^{-1}$. The layer has a temperature of 305.15 K, and the temperature of the body is assumed to be 310.15 K.

temperature than would be the case without a layer. The lossless case results in brightness temperatures which vary about the (near-constant) value when no layer is present; thus, for a lossless layer the presence of a layer may increase or decrease the observed brightness temperature of the target dependent on the chosen frequency.

6.7 Conclusion

Passive millimeter-wave sensors have been described for short-range detection of plastic and metal concealed objects. Both super heterodyne and direct detection versions operating in W-band were assembled and tested with commercial off-the-shelf components demonstrating the capability to detect various concealed objects with high spatial resolution and detection probability up to 100%. By equipping the sensor with antenna with higher gain and more narrow beams, the operating range can be extended by 15–20m. In this case, the sensor must be mounted on a tripod for manual control by the operator. Once installed on Az-El programmable positioner, the sensor can operate as an imaging scanner. Examples of this mode of operation were given for indoor and outdoor environments. Recent achievements in design and manufacturing W-band Silicon RFIC components [27] open new capabilities in miniaturizing the next generation of sensors.

References

1. B. Lee, *Theoretical prediction and measurement of the fabric surface apparent temperature in a simulated man/fabric/environment system*, Report DSTO-TR-0849, DSTO Aeronautical and Maritime Research Laboratory, Australia, 1999.
2. http://en.wikipedia.org/wiki/Black-body_radiation.
3. R. W. McMillan, N. C. Me, D. D. Ferris Jr., and M. C. Wicks, Concealed weapon detection using microwave and millimeter wave sensors, in *Proceedings of Microwave and Millimeter Wave Technology Conference, ICMMT '98*, 1998, pp. 1–4.
4. B. Kapilevich, B. Litvak, and A. Sulzinger, Einat: passive mm-wave detector, International Patent WO2012/140587 A2, 18 October 2012.
5. N. Skou and D. Le Vine, *Microwave radiometer systems: design and analysis*, Artech House, Norwood, MA, 1989.
6. R. H. Dicke, The measurement of thermal radiation at microwave frequencies, *Review of Scientific Instruments*, 17, 68–275, 1946.
7. Microwave and millimeter-wave radiometers, http://www.millitech.com/pdfs/Radiometer.pdf.
8. See http://www.elva-1.com.

9. B. Kapilevich, B. Litvak, and A. Shulzinger, Passive nonimaging mm-wave sensor for detecting hidden objects, in *IEEE International Conference on Microwaves, Communications, Antennas, and Electronic Systems (COMCAS 2013)*, Tel Aviv, Israel, 21–23 October 2013, pp. 1–5.
10. M. R. Fetterman, J. Grata, G. Jubic, W. L. Kiser Jr., and A. Visnansky, Simulation, acquisition and analysis of passive millimeter-wave images in remote sensing applications, *Optics Express*, 16(25), 20503–20515, 2008.
11. H. Sato, T. Hirose, T. Ohki, T. Takahashi, K. Makiyama, H. Kobayashi, K. Sawaya, and K. Mizuno, Compact receiver module for a 94 GHz band passive millimeter-wave imager, *IET Microwaves, Antennas and Propagation*, 2(8), 848–853, 2008.
12. L. S. Sadovnik, V. A. Manasson, R. E. Chapman, R. M. Mino, and V. Kiseliov, Remote fire detection using MMW radiometric sensor, in *Proceedings of SPIE 3378, Passive Millimeter-Wave Imaging Technology II*, 1998, pp. 73–80.
13. H. P. Moyer, J. H. Schaffner, Y. Royter, M. Sokolich, B. Hughes, Y. J. Yoon, and J. N. Schulman, Passive millimeter-wave imaging module with preamplified zero-bias detection, *IEEE Transactions on MTT*, 56(7), 1592–1600, 2008.
14. J. D. Kraus, *Radio astronomy*, McGraw Hill, New York, 1966.
15. F. T. Ulaby, R. K. Moore, and A. K. Fung, *Microwave remote sensing: active and passive*, vol. 1, Addison-Wesley, Reading, MA, 1981.
16. J. A. Nanzer and R. L. Rogers, Human presence detection using millimeter-wave radiometry, *IEEE Transactions on MTT*, 55(12), 2727–2733, 2007.
17. J. A. Nanzer and R. L. Rogers, Applying millimeter-wave radiometry to the detection of self-luminous objects at close range, *IEEE Transactions on MTT*, 56(9), 2054–2061, 2008.
18. J. A. Nanzer and R. L. Rogers, Analysis of the signal response of a scanning-beam millimeter-wave correlation radiometer, *IEEE Transactions on MTT*, 59(9), 2357–2368, 2011.
19. J. A. Nanzer, Millimeter-wave interferometric angular velocity detection, *IEEE Transactions on MTT*, 58(12), 4128–4136, 2010.
20. J. Polivka, An overview of microwave sensor technology, *High Frequency Electronics*, 32–42, 2007.
21. J. Polivka, Noise can be good, too, *Microwave Journal*, 47, 66–78, 2004.
22. J. Bodycomb, P. Helm, M. Kane, D. Nicol, and I. Ferguson, Thin film and emissivity effects on radiometric temperature measurement accuracy, in *11th IEEE International Conference on Advanced Thermal Processing of Semiconductors*, 2003, pp. 117–123.
23. R. Appleby and H. B. Wallace, Standoff detection of weapons and contraband in the 100 GHz to 1 THz region, *IEEE Transactions on Antennas and Propagation*, 55(11), 2944–2956, 2007.
24. C. M. Alabaster, Permittivity of human skin in the millimetre wave band, *IEEE Electronics Letters*, 39(21), 1521–1522, 2003.
25. B. Kapilevich, B. Litvak, M. Einat, O. Shotman, Passive mm-wave sensor for indoor and outdoor homeland security applications, in *Proceedings of the International Conference on Sensor Technologies and Applications*, Spain, 2007, pp. 20–23.
26. B. Kapilevich, B. Litvak, A. Shulzinger, and M. Einat, Portable passive millimetre-wave sensor for detecting concealed weapons and explosives hidden on a human body, *IEEE Sensors Journal*, 13, 11, 4224–4228, 2013.

27. O. Inac, F. Golcuk, T. Kanar, and G. M. Rebeiz, A 90–100 GHz phased-array transmit/receive silicon RFIC module with built-in self-test, *IEEE Transactions on Microwave Theory and Techniques*, 61, 10, 3774–3782, 2013.
28. B. Kapilevich, B. Litvak, O. Shotman, and M. Einat, Wide band mm-wave sensor: design and applications, in *IEEE 24th Convention of Electrical and Electronics Engineers in Israel*, 2006, pp. 161–164.

7

The Role of Shielding Effects in Operating Non-Imaging Sensors

7.1 Basic Shielding Mechanisms

The overall efficiency of detecting concealed objects is greatly dependent on the electromagnetic properties and physical structure of obscuring layers. The location of a hidden object with respect to some background material may be different, as shown in Figure 7.1:

- The object is located on a surface of a background material, for example, on a human body, and covered by obscuring materials, for example, textile or leather clothing (Figure 7.1a).
- The object is buried in a sub-surface position within the background material (Figure 7.1b).

If the object is located at the surface, the electromagnetic waves incident upon the target are partially reflected and scattered by any coating materials. If the optical thickness of this material is comparable with the wavelength of the incident radiation, then there may be significant alteration in reflectance and absorption of the incident radiation due to thin-film interference effects (see Section 6.5 for a discussion on thin-film interference effects). The texture of the coating material, such as woven fabric, also affects its reflectance/transmittance characteristics [1] and may lead to a Rayleigh scattering effect in the presence of a rough surface [2]. A similar situation takes place in the case of sub-surface position of the target.

The role of coating materials in shielding the target varies widely depending on the nature of the coating materials [3]. These effects can be accurately predicted with knowledge of the complex permittivity of the covering layer and its physical structure. Various modifications of the free space method [4] are widely used for millimetre wave and THz characterization of dielectric materials. An isotropic slab in air [5] is the simple theoretical model employed for retrieving the complex permittivity from the measured data. However, the approximation of textiles, which generally have an anisotropic

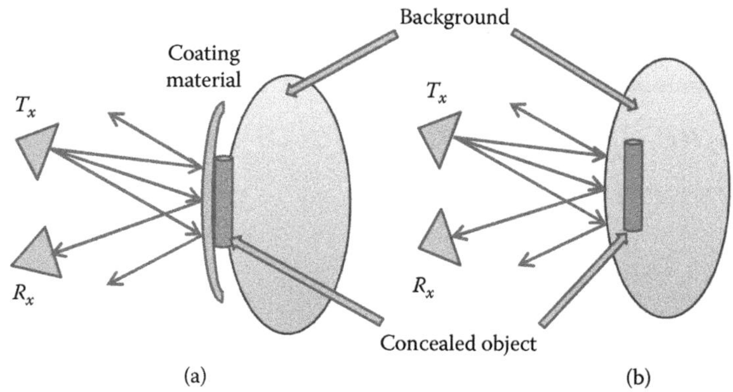

FIGURE 7.1
The location of the object with respect to a background material: (a) on the surface of the background and (b) the object is buried within a sub-surface.

structure, to a uniform dielectric slab is not necessarily a good one. Indeed, if texture or surface roughness is comparable with wavelength, Rayleigh scattering may drastically degrade the measurement accuracy. One way to mitigate the problem is the development of special data smoothing algorithms [6] needed to avoid undesirable rapid oscillations caused by parasitic standing wave interference between R_x and T_x antennae and the influence of other factors. As a result, the raw transmittance data cannot be used for a validation of the best-fitting permittivity without the additional specific processing.

Assuming that textiles and leathers are isotropic materials, they must have complex permittivity values which are constant, regardless of an incidence angle. Variations of the real and imaginary parts of the dielectric constant, as a function of the incident angle, can be considered as a measure of anisotropy and non-uniformity of the thickness of the sample. This parameter has a statistical nature and should be characterized by standard deviation of the measured material characteristics in the angular domain.

The free space approach is most suitable in the mm-wave range since the sample size has reasonable dimensions in order to avoid undesirable edge effects. However, the drawback of most of the free space measuring systems is the presence of ripples (or parasitic resonances) in the measured data caused by multiple reflections from sample surfaces and antennae. These effects are not easily predicted, and thus mitigated, by theoretical models. As a result, various smoothing algorithms based on heuristic criteria may be used to reduce degradation of measurement accuracy [8], [9].

In order to minimize parasitic reflectance, a matched mm-wave/THz power metre can be used as reported in [10]. This power metre has an absorbing thin-film radio frequency (RF) sensor operating near the Brewster angle, resulting in a minimum reflectance at the input of the power metre. The complex dielectric constant is derived from the angular dependence of the measured power transmittance using the root-finding procedure applied to the proper formulated system of non-linear equations. A choice of starting values providing the unique solution must be properly chosen in order to avoid a divergence of the root-finding process.

7.2 Theoretical Models and Estimation Method

7.2.1 Basic Existing Models

Assume that an infinite dielectric slab of thickness d which is illuminated by plane wave with an incident angle θ relative to the normal of the slab. The reflection R and transmission T coefficients, being basic physical parameters measured in the free space methods, can be presented in general form as non-linear functions of d, θ, frequency f and the relative complex dielectric constant $\varepsilon_r = \varepsilon_r' - j\,\varepsilon_r''$:

$$R = r(f, d, \theta, \varepsilon_r', \varepsilon_r'')$$

$$T = t(f, d, \theta, \varepsilon_r', \varepsilon_r'') \tag{7.1}$$

If an experimental setup is equipped by a vector network analyzer (VNA), the magnitude and phase of R and T can be measured and the complex dielectric constant is retrieved from Equation 7.1 using a standard root-finding procedure. However, VNAs operating in sub-mm and THz bands are very expensive. Alternatively, an amplitude-only transmittance can be measured quite easily without recourse to expensive test equipment [14], [17]. The models, corresponding to this amplitude-only approach are considered below.

Frequency domain formulation: The R and T magnitudes are measured at fixed incident angle θ, within the specified frequency range subdivided on M frequency points. Then, the value functions V_{Rf} and V_{Tf} are formulated as follows:

$$V_{Rf} = \frac{1}{M}\sum_{m=1}^{M}\left[|R_m| - |r(f_m, d, \theta, \varepsilon_r', \varepsilon_r'')|\right]^2 \tag{7.2}$$

$$V_{Tf} = \frac{1}{M}\sum_{m=1}^{M}\left[|T_m| - |t(f_m, d, \theta, \varepsilon_r', \varepsilon_r'')|\right]^2 \tag{7.3}$$

The unknown dielectric constant $\varepsilon_r = \varepsilon_r'' - j\,\varepsilon_r''$ are determined by minimization (Equation 7.2) and (Equation 7. 3) in $\{\varepsilon_r', \varepsilon_r''\}$ space using a non-linear regression algorithm, assuming that ε_r is frequency independent over the frequency range measured. The number of frequency points M depends on hardware facilities and the chosen algorithm's efficiency. A typical range for M is ~ 100–1000. As a result of minimization, the best values of ε_r' and ε_r'' satisfying a least-squares criterion are determined. Examples of measurements based on the frequency domain formulation are reported in [11].

Angular domain formulation: The R and T magnitudes are measured at fixed frequency f within the specified incident angle range at N angular points. Then, the value functions $V_{R\theta}$ and $V_{T\theta}$ are formulated as follows:

$$V_{R\theta} = \frac{1}{N}\sum_{n=1}^{N}\left[\left|R_n\right| - \left|r\left(f, d, \theta_n, \varepsilon_r', \varepsilon_r''\right)\right|\right]^2 \tag{7.4}$$

$$V_{T\theta} = \frac{1}{N}\sum_{n=1}^{N}\left[\left|T_n\right| - \left|t\left(f, d, \theta_n, \varepsilon_r', \varepsilon_r''\right)\right|\right]^2 \tag{7.5}$$

As a result of non-linear optimization the best values of ε_r' and ε_r'' satisfying a least-squares criterion can be restored. Examples of measurements based on the angular domain formulation were reported in [12], [13], [34].

Space division formulation: If two slabs of the same material and thicknesses d_1 and d_2 are separated by the distance L, a structure similar to a Fabry-Perot resonator can be created. Variation of a distance between slabs forms the transmittance interferogram pattern that can be used for retreiving ε_r' and ε_r''. The R and T magnitudes are measured at the fixed frequency f and incident angle θ (typically $\theta = 0°$) within P longitudinal points. Then, the value functions V_{RL} and V_{TL} are formulated as follows:

$$V_{RL} = \frac{1}{P}\sum_{p=1}^{P}\left[\left|R_p\right| - \left|r\left(f, d_1, d_2, \theta, L_p, \varepsilon_r', \varepsilon_r''\right)\right|\right]^2 \tag{7.6}$$

$$V_{TL} = \frac{1}{P}\sum_{p=1}^{P}\left[\left|T_p\right| - \left|t\left(f, d_1, d_2, \theta, L_p, \varepsilon_r', \varepsilon_r''\right)\right|\right]^2 \tag{7.7}$$

Again, as a result of non-linear optimization the best values of ε_r' and ε_r'' satisfying a least-squares criterion are restored. Examples of measurements based on the space division formulation were reported in [14], [15].

It should be pointed out the problem of obtaining the complex permittivity using any of the mentioned techniques is an ill-conditioned one. Degeneracy and divergence are typical for all the above formulations. They are best solved by choosing initial parameters (guess values) that are not too far from the true solution. Although this will be a subjective judgement, in many practical situations it is sufficient to find a good starting point for the non-linear refinement. However, evaluation of R_p and T_p needed to reach such a refinement must be made accurately in order to avoid spurious results. Multiple reflections, which are inherent to most free space realizations, are a serious source of measurement inaccuracy in all the above approaches. So, the technique providing a minimization of such reflections is strongly preferable.

7.2.2 Estimation Method Employed

One of the ways to avoid multiple reflections is to employ the matched power metre (MPM). It provides direct estimate of RF power delivered to its input and demonstrates well-matched operation if an incident angle at the input port of MPM is near the Brewster angle of the absorbing thin film. Due to higher accuracy in determination of incident RF power with a MPM, it is possible to replace least-squares minimization [16] by the numerical root-finding algorithm applied to the system of non-linear equations [10]:

$$\begin{aligned} \left|t(f,d,\theta_1,\varepsilon_r',\varepsilon_r'')\right|^2 &= T_{p1} \\ \left|t(f,d,\theta_2,\varepsilon_r',\varepsilon_r'')\right|^2 &= T_{p2} \end{aligned} \tag{7.8}$$

where T_{p1} and T_{p2} are power transmission coefficients measured at the incident angles θ_1 and θ_2, respectively.

The theoretical expression of the transmittance of the dielectric slab shown in Figure 7.2 can be derived using boundary conditions [18, 35].

According to [5] the power reflection and transmission coefficients of the slab sample are given by the expression

$$R_{s,p} = \left|\Gamma_{s,p}\right|^2$$

$$T_{s,p} = \left|\frac{\left(1-\Gamma_{s,p}^2\right)\exp(i\kappa d)}{1-\Gamma_{s,p}^2\exp(i2\kappa d)}\right|^2 \tag{7.9}$$

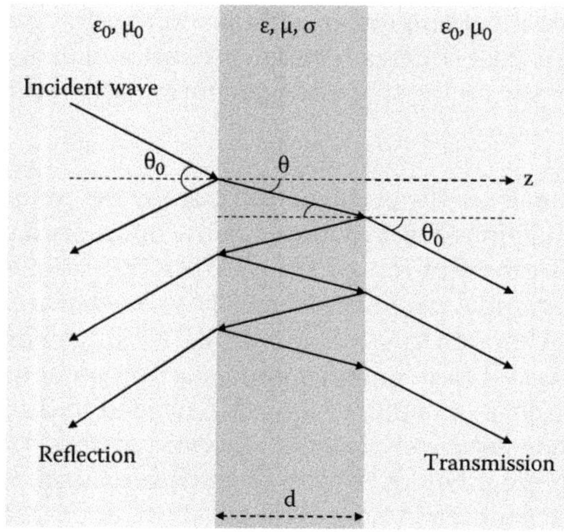

FIGURE 7.2
Dielectric slab used in modelling.

$$\kappa = \frac{2\pi}{\lambda}\sqrt{\varepsilon_r - \sin^2\theta}$$

$$\Gamma_s = \frac{\cos\theta - \sqrt{\varepsilon_r - \sin^2\theta}}{\cos\theta + \sqrt{\varepsilon_r - \sin^2\theta}}$$

$$\Gamma_p = \frac{\varepsilon_r\cos\theta - \sqrt{\varepsilon_r - \sin^2\theta}}{\varepsilon_r\cos\theta + \sqrt{\varepsilon_r - \sin^2\theta}}$$

7.3 Well/Ill-Conditioning Analysis

The existence of multiple solutions of the system of non-linear equations (7.8) needs well/ill-conditioning analysis of power transmittance in space $\{\varepsilon_r', \varepsilon_r''\}$. In order to avoid spurious roots of Equation 7.8 or a diverging solution, it is necessary to guess "good" initial values for an unknown ε_r' and ε_r''. Preliminary analysis of well/ill-conditioning behaviour in $\{\varepsilon_r', \varepsilon_r''\}$ space permits determination of the relevant starting point in searching for a solution of Equation 7.8. As an example, consider a surface plot of power transmission coefficient T_p in $\{\varepsilon_r', \varepsilon_r''\}$ space for different slab thicknesses and incident angles, assuming that f is 330 GHz. Several such surfaces of T_p reconstructed from Equation 7.9 are

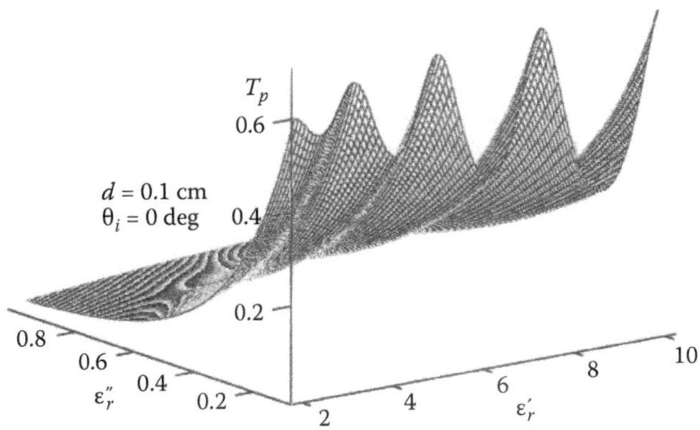

FIGURE 7.3
A surface plot of power transmission coefficient T_p in the $\{\varepsilon_r', \varepsilon_r''\}$ space for incident angle θ_i = 0° and slab thickness d = 0.1 cm. (From Kapilevich, B., Pinhasi, Y. and Litvak, B., *International Journal of Infrared, Millimeter and Terahertz Waves*, 32, 1446–1456, 2011. With kind permission of Springer Science+Business Media.)

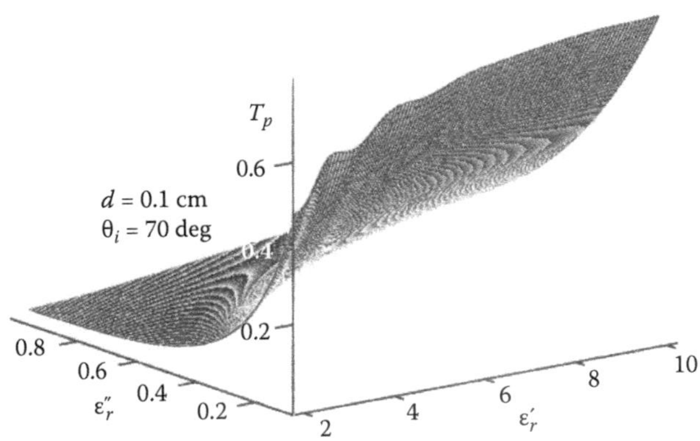

FIGURE 7.4
A surface plot of power transmission coefficient T_p in the $\{\varepsilon_r', \varepsilon_r''\}$ space for incident angle θ_i = 70° and slab thickness d = 0.1 cm. (From Kapilevich, B., Pinhasi, Y. and Litvak, B., *International Journal of Infrared, Millimeter and Terahertz Waves*, 32, 1446–1456, 2011. With kind permission of Springer Science+Business Media.)

depicted in Figures 7.3–7.6 for the slab thicknesses d = 0.1 cm and 0.25 cm and for incident angles θ_i = 0° and θ_i = 70°. Comparing these figures reveals that quite smooth surfaces corresponding to well-conditioned areas exist within smaller values of ε_r' and relatively large values of ε_r''. This area tends to widen with increasing incident angle θ_i and thickness d. In addition, quasi-periodical variations of T_p, typical for ill-conditioned behaviour, become more significant with increasing slab thickness for smaller ε_r''. So, it is recommended to use a starting solution for Equation 7.8 that is of higher ε_r'' (lossy material) and smaller ε_r'.

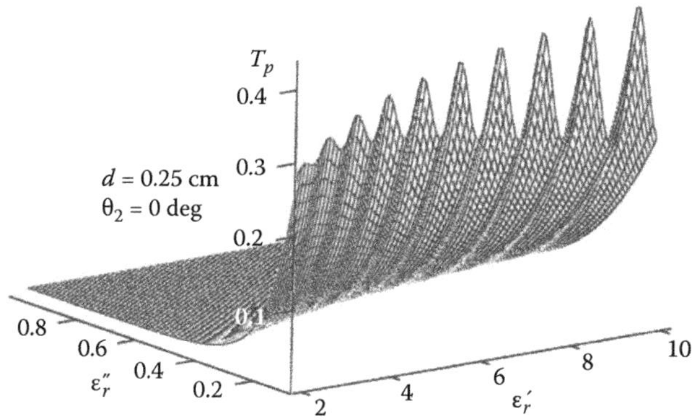

FIGURE 7.5
A surface plot of power transmission coefficient T_p in the $\{\varepsilon_r', \varepsilon_r''\}$ space for incident angle θ_i = 0° and slab thickness d = 0.25 cm. (From Kapilevich, B., Pinhasi, Y. and Litvak, B., *International Journal of Infrared, Millimeter and Terahertz Waves*, 32, 1446–1456, 2011. With kind permission of Springer Science+Business Media.)

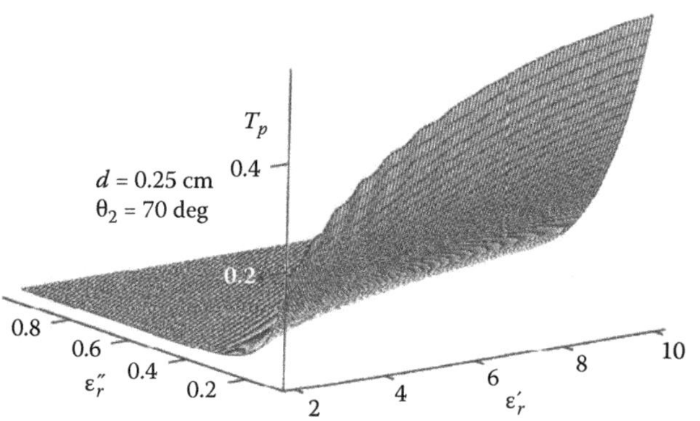

FIGURE 7.6
A surface plot of power transmission coefficient T_p in the $\{\varepsilon_r', \varepsilon_r''\}$ space for incident angle θ_i = 70° and slab thickness d = 0.25 cm. (From Kapilevich, B., Pinhasi, Y. and Litvak, B., *International Journal of Infrared, Millimeter and Terahertz Waves*, 32, 1446–1456, 2011. With kind permission of Springer Science+Business Media.)

7.4 Experimental Setup and Results of Measurements: Coherent Illumination

Figure 7.7a shows schematically the experimental setup assembled for material characterization near f = 330 GHz. It consists of an X-band synthesizer with a power amplifier at its ouput that drives a multi-stage multiplier (four sections) with the total multiplication factor 32 = 4 × 2 × 2 × 2. The output signal, about 10 mW, is radiated by the horn-lens antenna (1.3° beamwidth and

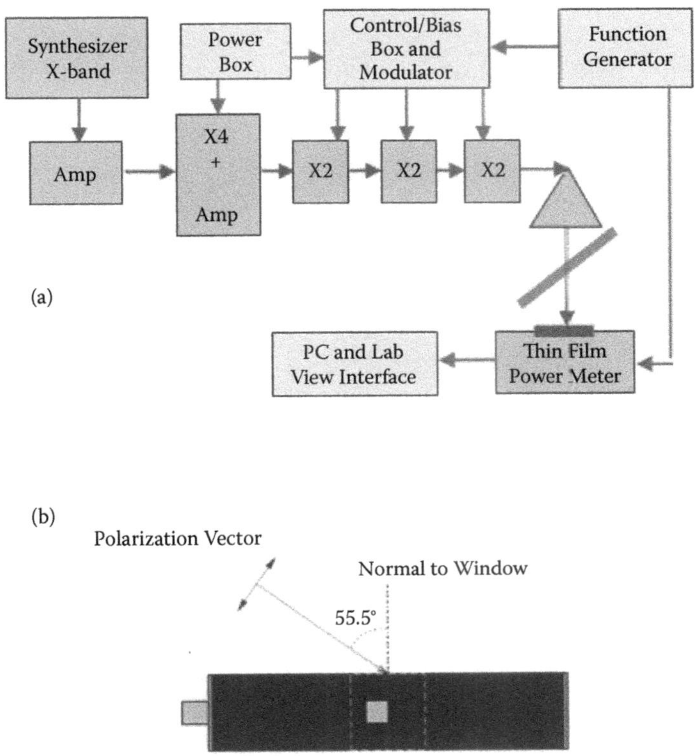

FIGURE 7.7
(a) The experimental setup and (b) the position the head aligned at the Brewster angle to the incoming beam in order to minimize reflections from the input window. (From Kapilevich, B. et al., in *IEEE International Conference on Microwaves, Communications, Antennas and Electronics Systems (COMCAS)*, Tel Aviv, Israel, 1–4, 2011. © 2011 by the Institute of Electrical and Electronics Engineers [IEEE].)

42 dBi gain). Energy propagating through the dielectric slab is absorbed by the thin-film element of the MPM (Thomas Keating THz power metre, http://www.terahertz.co.uk/). The input window of the MPM must be aligned at the Brewster angle as shown in Figure 7.7b in order to reach the best matching condition. The power box, the control bias box and the modulator provide proper functionality of each element of the setup, while a function generator is employed for synchronization purposes. Figure 7.8 illustrates a general photo of this setup.

As an example, the power transmisson coefficients have been measured for a sample of ceramic tile of thickness 0.25 cm and a lossy plastic sheet of thickness 1.21 cm. Corresponding measured data are depicted in Figure 7.9 by stepped lines. The entry data used for determination of ε_r' and ε_r'' are summarized in Table 7.1. Comparison of several methods of numerical solution of Equation 7.8 has revealed the quasi-Newton method [16] is superior in terms of the stability and convergence for a wider range of inital guess values.

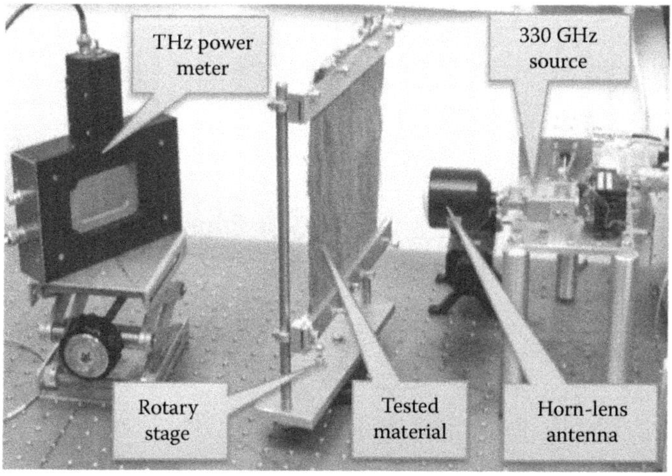

FIGURE 7.8 (See color insert.)
Photograph of the experimental setup. From Kapilevich, B., Litvak, B., Anisimov, M., Hardon, D. and Pinhasi, Y., *International Journal of Microwave Science and Technology*, 2012, doi:10.1155/2012/375601. Reproduced courtesy of Hindawi Publishing Corp.)

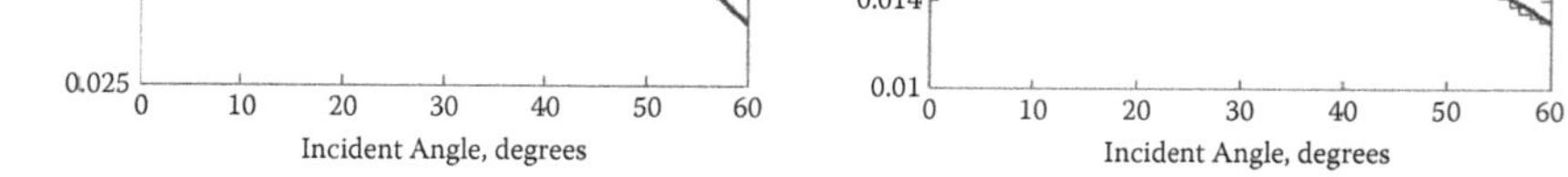

FIGURE 7.9
The measured (stepped line) and calculated (solid line) power transmisson coefficients for the sample of the ceramic tile of thickness 0.25 cm and lossy plastic sheet of the thickness 1.21 cm. (From Kapilevich, B. et al., in *IEEE International Conference on Microwaves, Communications, Antennas and Electronics Systems (COMCAS)*, Tel Aviv, Israel, 1–4, 2011. © 2011 by the Institute of Electrical and Electronics Engineers [IEEE].)

The relatively wide range of starting variations proves the well-conditioned behaviour of the value function for materials under examination.

In order to evaluate a validity of the root-finding procedure, the transmission coefficients were re-calculated using Equation 7.9 with the values of dielectric constants determined from the measurements. The results of such re-calculations are shown in Figure 7.9 by solid lines. One may conclude that the measured and re-calculated transmittances are in close agreement, proving the validity of the root-finding method and providing an acceptable measurement accuracy.

TABLE 7.1

Summary of Entry Data Used for Numerical Solution System (Equation 7.8)

Parameter \ Material	Ceramic Tile $d = 0.25$ cm	Lossy Plastic Sheet $d = 1.21$ cm
Incident angle θ_1	10°	10°
Incident angle θ_2	50°	50°
Power transmission coefficient at θ_1	0.037	0.024
Power transmission coefficient at θ_2	0.03	0.0157
Calculated real part of dielectric constant, ε_r'	3.03	2.44
Calculated imaginary part of dielectric constant, ε_r''	0.316	0.067
Range of initial guess values for real part of dielectric constant	$1 < \varepsilon_r' < 20$	$1 < \varepsilon_r' < 10$
Range of initial guess values for imaginary part of dielectric constant	$0 < \varepsilon_r'' < 3$	$0 < \varepsilon_r'' < 0.6$

Several types of textiles and leathers have been tested using the outlined method, at frequencies of 330, 220 and 110 GHz. The results for these materials are summarized in Tables 7.2 and 7.3 for the incident angle of 0°.

In order to simplify the process of retrieving the complex permittivity from the measured transmittance, the two MATLAB® graphic user interfaces (GUIs) have been developed.

GUI-1 consists of the two blocks (see Figure 7.10):

1. A block for the input data:
   - *d*—the thickness of the dielectric slab, cm
   - The two incidence angles *theta*0 and *theta*1 in degrees
   - The measured power transmittances $|T_0|$ and $|T_1|$ corresponding to incidence angles *theta*0 and *theta*1
   - The initial guess value of complex dielectric constant ε_{r0}
2. A block for the evaluated result:
   - An evaluate button activating restoration of ε_r
   - A window showing the calculated complex permittivity ε_r for the input entry data as a result of numerical solution of the system of non-linear equations (7.8).

GUI-2 consists of the two blocks (see Figure 7.11):

1. The block for input data:
   - *f*—the operating frequency, GHz
   - *d*—the thickness of the dielectric slab, cm

TABLE 7.2

Table of Textile Samples under Test

Material	Sample #1 (denim)	Sample #2	Sample #3	Sample #4	Sample #5	Sample #6 (wool scarf)	Sample #7 (stockinet)	Sample #8 (satin)
General view of samples under testing							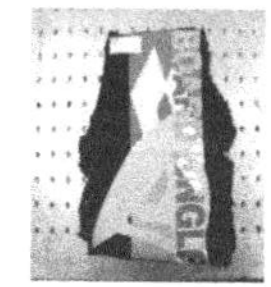	
Thickness d, mm	0.8	0.45	0.25	0.7	0.25	1.6	0.65	0.2
Attenuation at normal incidence, dB/mm at 330 GHz	−1.42	−1.34	−1.45	−0.94	−2.06	−2.85	−2.67	−2.045
Attenuation at normal incidence, dB/mm at 220 GHz	−0.5720	−0.4950	−1.0749	−0.1253	−1.4485	−0.9335	−1.9221	−0.8864
Attenuation at normal incidence, dB/mm at 110 GHz	−0.9465	−1.1247	−2.4192	−0.2533	−3.2369	−0.9296	−2.1021	−0.5275
ε_r at 330 GHz	2.73	2.74	2.72	3.66	2.54	3.18	3.22	2.74
$\tan\delta$ at 330 GHz	0.027	0.011	0.025	0.011	0.026	0.047	0.025	0.027
ε_r at 220 GHz	2.496	1.993	2.29	3.313	2.277	2.639	2.876	2.475
$\tan\delta$ at 220 GHz	0.018	0.02	0.013	0.003	0.014	0.026	0.038	0.019
ε_r at 110 GHz	2.92	3.04	1.91	3.02	2.14	2.16	2.49	2.39
$\tan\delta$ at 110 GHz	0.029	0.03	0.021	0.009	0.03	0.035	0.043	0.012

Source: Kapilevich, B., Litvak, B., Anisimov, M., Hardon, D. and Pinhasi, Y., *International Journal of Microwave Science and Technology*, vol. 2012, Article ID 375601, 2012, doi:10.1155/2012/375601. Reproduced courtesy of Hindawi Publishing Corp.

TABLE 7.3

Table of Leather Samples under Test

Material	Sample #0 (natural leather)	Sample #1 (artificial leather)	Sample #2 (artificial leather)	Sample #3 (artificial leather)	Sample #4 (artificial leather)	Sample #5 (artificial leather)
General view of samples under testing			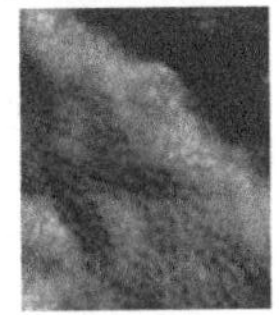			
Thickness d, mm	1.15	0.95	1	0.6	0.8	0.6
Attenuation at normal incidence, dB/mm at 330 GHz	–2.77	–2.11	–1.67	–1.48	–1.86	–1.89
Attenuation at normal incidence, dB/mm at 220 GHz	–0.64	–0.24	–1.9	–1.95	–1.16	–2.73
Attenuation at normal incidence, dB/mm at 110 GHz	–0.66	–0.14	–0.32	–1.1	–0.28	–2.41
ε_r at 330 GHz	3.4	3.104	3.008	2.17	3.02	2.49
tanδ at 330 GHz	0.127	0.079	0.08	0.066	0.084	0.085
ε_r at 220 GHz	3.93	2.06	2.4	1.5	2.09	1.9
tanδ at 220 GHz	0.023	0.025	0.015	0.034	0.026	0.027
ε_r at 110 GHz		1.92	1.9	2.27	2.6	1.85
tanδ at 110 GHz	0.018	0.025	0.022	0.057	0.012	0.045

Source: Kapilevich, B., Litvak, B., Anisimov, M., Hardon, D. and Pinhasi, Y., *International Journal of Microwave Science and Technology*, vol. 2012, Article ID 375601, 2012, doi:10.1155/2012/375601. Reproduced courtesy of Hindawi Publishing Corp.

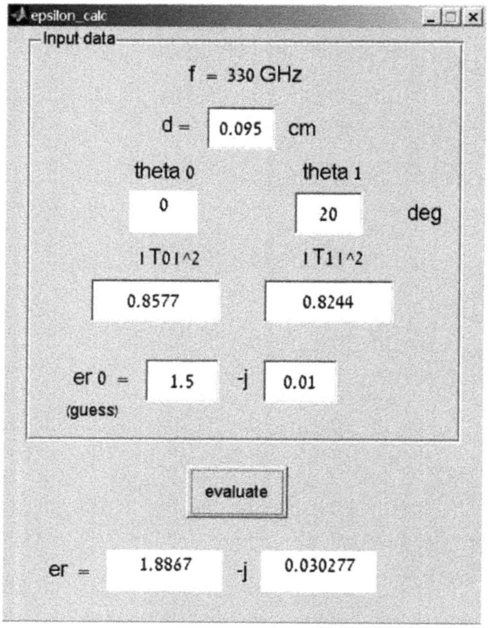

FIGURE 7.10
A screenshot of GUI-1 used for calculations of complex permittivity from GUI-2.

- The incidence angle *theta* (single evaluation) in degrees
- The array of the incidence angles *theta* in degrees
- The complex dielectric constant ε_r

2. The block of the evaluated result:
   - The evaluate button for single calculation
   - The window showing the calculated power transmittance from Equation 7.9
   - The evaluate button for drawing the plot—the power transmittance as a function of an incidence angle

Figures 7.12 and 7.13 show the measured transmittances as a function of incidence angle for textile and leather samples, respectively. The polarization is matched with the polarization of the MPM. The reconstruction procedure based on transmittance measurements has been employed.

The solution of the system of non-linear equations (7.8) requires knowledge of the two independent transmittances for specified incidence angles. They can be chosen arbitrarily, but we have fixed one of them assuming that $\theta_1 = 0$, while the second angle was varied over the range $5° \le \theta_2 \le 50°$. Tables 7.4 and 7.5 summarize the retrieved real and imaginary parts of dielectric constants

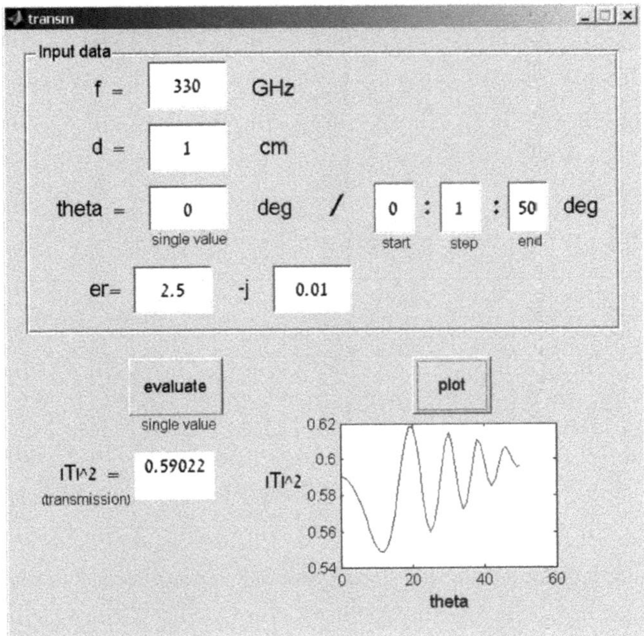

FIGURE 7.11
The screenshot of GUI-2 used for calculations of power transmittance.

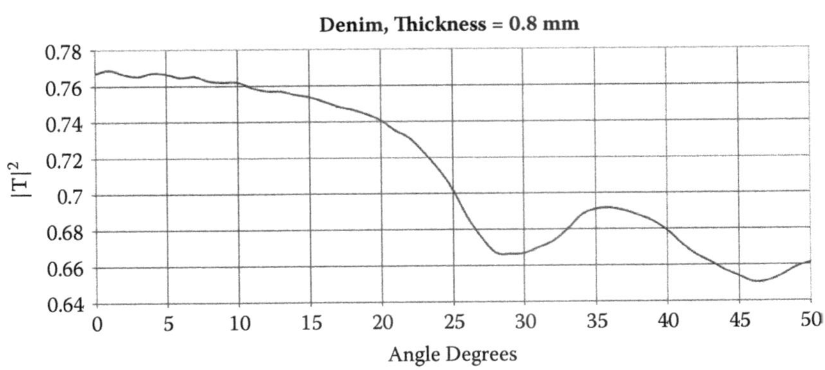

FIGURE 7.12
Measured power transmittance of textile samples as a function of incidence angle.

of the textile samples for different incidence angles. Similar data are given in Tables 7.6 and 7.7 for leather samples.

Both textile and leather samples have demonstrated variations of complex permittivity for different incidence angles due to non-uniform thickness, surface roughness, specific texture and non-ideal flatness. All these factors

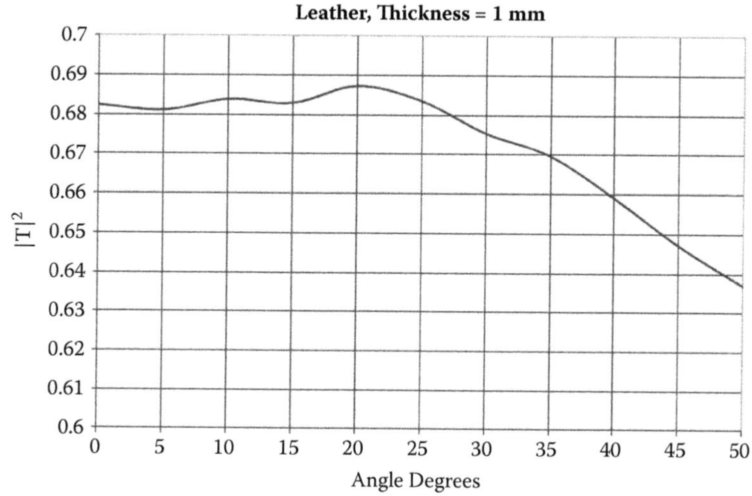

FIGURE 7.13
Measured power transmittance of leather samples as a function of incidence angle.

TABLE 7.4

Real Part of Permittivity Evaluated for the Textile Samples: $\theta_1 = 0$ and Different Incidence Angles θ_2

Incidence Angle, θ_2 Degrees	Sample 1 (denim)	Sample 2	Sample 3	Sample 4	Sample 5	Sample 6 (wool scarf)	Sample 7 (stockinette)	Sample 8 (satin)
5	2.83	2.77	2.73	3.65	2.54	3.2	3.15	2.73
10	2.71	2.71	2.7	3.76	2.43	3.21	3.13	2.69
15	2.67	2.92	2.74	3.67	2.41	3.23	3.17	2.73
20	2.7	2.76	2.77	3.67	2.49	3.12	3.17	2.75
25	2.73	2.68	2.79	3.65	2.51	3.11	3.2	2.76
30	2.69	2.5	2.75	3.67	2.54	3.14	3.34	2.75
35	2.67	2.26	2.7	3.66	2.54	3.17	3.34	2.76
40	2.75	2.84	2.75	3.68	2.58	3.2	3.22	2.76
45	2.78	2.95	2.59	3.66	2.68	3.22	3.26	2.75
50	2.79	3.01	—	3.58	2.68	3.23	3.21	2.76
Averaged, ε_r'	2.73	2.74	2.72	3.66	2.54	3.18	3.22	2.74
Standard deviation	0.054	0.253	0.0623	0.0458	0.0956	0.0452	0.0737	0.0216

Source: Kapilevich, B., Litvak, B., Anisimov, M., Hardon, D. and Pinhasi, Y., *International Journal of Microwave Science and Technology*, vol. 2012, Article ID 375601, 2012, doi:10.1155/2012/375601. Reproduced courtesy of Hindawi Publishing Corp.

TABLE 7.5

Imaginary Part of Permittivity Evaluated for the Textile Samples: $\theta_1 = 0$ and Different Incidence Angles θ_2

Incidence Angle, θ_2 Degrees	Sample 1 (cotton jeans)	Sample 2	Sample 3	Sample 4	Sample 5	Sample 6 (wool scarf)	Sample 7 (stockinette)	Sample 8 (satin)
5	0.0715	0.033	0.07	0.041	0.065	0.133	0.054	0.074
10	0.0633	0.029	0.057	0.028	0.06	0.13	0.057	0.072
15	0.0603	0.0103	0.06	0.038	0.045	0.134	0.041	0.071
20	0.0634	0.033	0.062	0.043	0.048	0.145	0.042	0.072
25	0.071	0.0397	0.065	0.044	0.052	0.158	0.054	0.072
30	0.0781	0.011	0.068	0.044	0.061	0.159	0.076	0.073
35	0.074	0.011	0.071	0.044	0.072	0.159	0.1	0.074
40	0.079	0.019	0.08	0.043	0.077	0.161	0.119	0.077
45	0.0858	0.06	0.085	0.044	0.089	0.172	0.126	0.08
50	0.0866	0.065	—	0.047	0.092	0.175	0.13	0.086
Averaged, ε_r''	0.073	0.031	0.068	0.042	0.066	0.15	0.08	0.075
Standard deviation	0.0092	0.0222	0.0097	0.0056	0.0172	0.0162	0.0356	0.0049

Source: Kapilevich, B., Litvak, B., Anisimov, M., Hardon, D. and Pinhasi, Y., *International Journal of Microwave Science and Technology*, vol. 2012, Article ID 375601, 2012, doi:10.1155/2012/375601. Reproduced courtesy of Hindawi Publishing Corp.

have a statistical nature and can be characterized by standard deviation (STD) presented in the last line of Tables 7.4–7.7.

In order to estimate the angular-invariant limits, it is assumed that the surface roughness satisfies the Rayleigh scattering condition [19]:

$$\Delta h < \frac{\lambda}{8\cos\theta_i} \tag{7.10}$$

where Δh is roughness variation, θ_i is incidence angle and λ is free space wavelength. Figure 7.14 illustrates schematically forward-scattered rays caused by surface roughness.

Since the roughness Δh and incidence angle θ_i are interrelated due to Equation 7.10, we can expect that the uncertainty in a determination of the complex permittivity should be increased with increasing θ_i too. We may ask, what is the acceptable range of incidence angles θ_i that give invariant complex permittivity values for the material under test? To answer the question, we need to compare the measured transmittance of a real material having rough surface with the calculated transmittance of the smoothed dielectric slab made from the same material. The results of such a comparison are shown in Figure 7.15a (textile sample 3, $\varepsilon_r' = 2.72$ and $\varepsilon_r'' = 0.068$) and Figure 7.15b (leather sample 0, $\varepsilon_r' = 3.4$ and $\varepsilon_r'' = 0.127$), where the lines marked by the symbol □ □ □ correspond to the measured transmittance, while the

TABLE 7.6

Real Part of Permittivity Evaluated for Leather Samples: $\theta_1 = 0$ and Different Incidence Angles θ_2

Incidence Angle, θ_2 Degrees	Sample 0	Sample 1	Sample 2	Sample 3	Sample 4	Sample 5
5	2.91	3.3	2.95	2.18	2.7	2.48
10	3.31	2.97	3.03	2.03	2.71	2.75
15	3.39	3.17	3.03	2.05	2.73	2.66
20	3.42	2.99	3.08	2.16	2.96	2.51
25	3.45	3.02	3.04	2.11	3.03	2.33
30	3.44	3.03	3.05	2.26	3.09	2.45
35	3.46	3.06	3.05	2.28	3.19	2.45
40	3.5	3.12	3.03	2.28	3.23	2.43
45	3.58	3.17	2.9	2.26	3.25	2.43
50	3.54	3.21	2.92	2.13	3.28	2.45
Averaged, ε_r'	3.4	3.104	3.008	2.17	3.02	2.49
Standard deviation	0.178	0.102	0.058	0.094	0.220	0.1221

Source: Kapilevich, B., Litvak, B., Anisimov, M., Hardon, D. and Pinhasi, Y., *International Journal of Microwave Science and Technology*, vol. 2012, Article ID 375601, 2012, doi:10.1155/2012/375601. Reproduced courtesy of Hindawi Publishing Corp.

TABLE 7.7

Imaginary Part of Permittivity for Leather Samples: $\theta_1 = 0$ and Different Incidence Angles θ_2

Incidence Angle, θ_2 Degrees	Sample 0	Sample 1	Sample 2	Sample 3	Sample 4	Sample 5
5	0.118	0.093	0.078	0.066	0.088	0.084
10	0.109	0.072	0.08	0.062	0.086	0.065
15	0.115	0.079	0.081	0.063	0.083	0.075
20	0.118	0.061	0.08	0.066	0.096	0.09
25	0.12	0.067	0.079	0.065	0.096	0.088
30	0.125	0.071	0.079	0.069	0.091	0.088
35	0.131	0.077	0.078	0.069	0.084	0.087
40	0.14	0.083	0.079	0.067	0.076	0.088
45	0.148	0.089	0.079	0.0676	0.0751	0.088
50	0.15	0.098	0.083	0.067	0.069	0.095
Averaged, ε_r''	0.127	0.079	0.08	0.066	0.084	0.085
Standard deviation	0.0135	0.0111	0.0014	0.0023	0.0086	0.0084

Source: Kapilevich, B., Litvak, B., Anisimov, M., Hardon, D. and Pinhasi, Y., *International Journal of Microwave Science and Technology*, vol. 2012, Article ID 375601, 2012, doi:10.1155/2012/375601. Reproduced courtesy of Hindawi Publishing Corp.

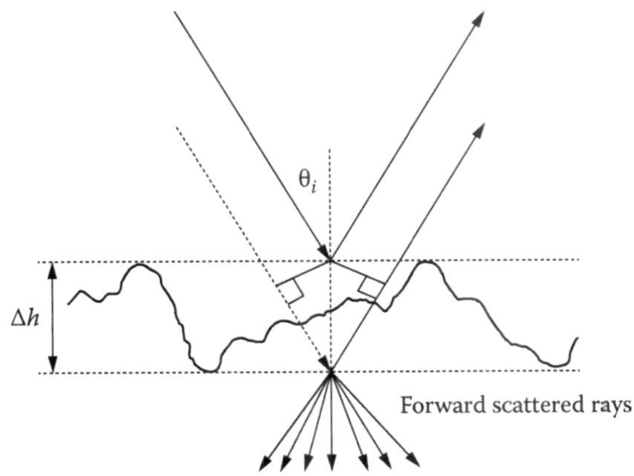

FIGURE 7.14
Generating forward scattered rays due to surface roughness. (From Kapilevich, B., Litvak, B., Anisimov, M., Hardon, D. and Pinhasi, Y., *International Journal of Microwave Science and Technology*, vol. 2012, Article ID 375601, 2012, doi:10.1155/2012/375601. Reproduced courtesy of Hindawi Publishing Corp.)

solid lines depict the calculated parameters. The averaged values of the real and imaginary parts of the dielectric constants, given in Tables 7.5–7.8, were employed in the theoretical predictions. The comparison between theory and experiment reveals the existence of a critical incidence angle, θ_c, that is determined by the angular invariance limit of the forward Rayleigh scattering as well as the incident wavelength. For the materials discussed, the critical angle is about 25° for leather samples and 30° for textile samples. If an incidence angle is greater than this critical value ($\theta_i > \theta_c$), the measurement of the bulk permittivity will be problematic. There is also a critical roughness variation Δh. Using Equation 7.10, this parameter was found to equal about 0.15 mm for the frequency of 330 GHz and $\theta_c = 30°$. The latter is a realistic estimate for the surface roughness of typical textiles and leather samples.

It is also important to formulate the angular-invariant limit from a theoretical model. The model must unify both deterministic and statistical properties of the real structure of materials under test. Microwave properties of weave architectures used in woven fabric composites have been investigated in [1] using a deterministic approach. Another approach is an application of the concept of the fractional Brownian motion appropriately describing natural surfaces [20] assuming that surface profiles can be approximated by the Weierstrass-Mandelbrot fractal functions. However, some entry parameters needed for real-world calculations such as the vertical height profile scaling factor, random variables which account for the amplitude and the phase behaviour of each tone, etc. [21], must be specified for materials under test and that is not a trivial problem.

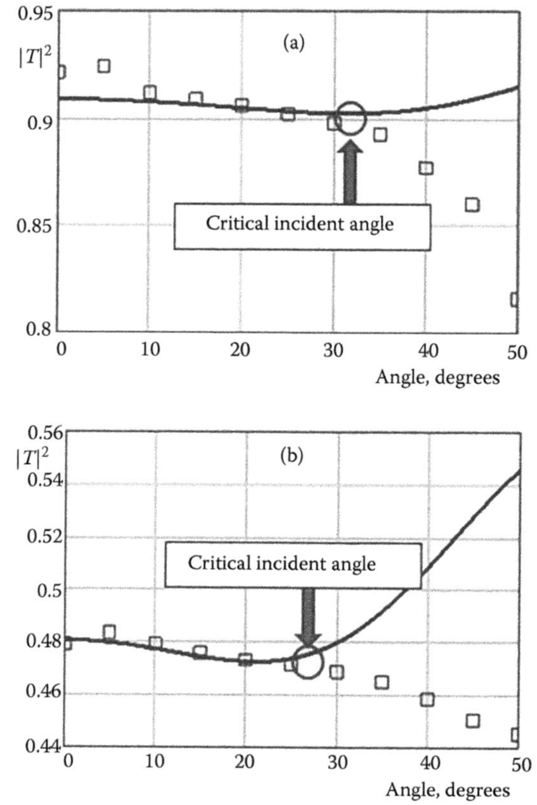

FIGURE 7.15
Comparison of calculated (solid line) and measured (symbol □ □ □) transmittances as a function of an incidence angle ((a) the textile sample #3, $\varepsilon_r' = 2.72$ and $\varepsilon_r'' = 0.068$); (b) the leather sample #0, $\varepsilon_r' = 3.4$ and $\varepsilon_r'' = 0.127$). (From Kapilevich, B., Litvak, B., Anisimov, M., Hardon, D. and Pinhasi, Y., *International Journal of Microwave Science and Technology*, vol. 2012, Article ID 375601, 2012, doi:10.1155/2012/375601. Reproduced courtesy of Hindawi Publishing Corp.)

TABLE 7.8
Summary of the Error Analysis Caused by the Drift of Measured RF Power

Parameter		
Incident angle	$\theta_1 = 10°$	$\theta_2 = 50°$
Power transmission coefficients	0.036 0.029	0.038 0.031
Range of the measured real part of dielectric constant, ε_r'	3.01–3.05 Error = 1.3%	
Range of the measured imaginary part of dielectric constant, ε_r''	0.314–0.317 Error = 3%	

7.5 Sensitivity and Error Analysis

The standard way for analysis of the sensitivity S of the measurements is a calculation of the proper partial derivatives, respectively, to the quantity of interest. We have performed these estimates for ε_r' and ε_r'', assuming that other quantities are fixed:

$$S = \begin{cases} \left.\dfrac{d|T_P|}{d(\varepsilon_r')}\right|_{f,d,\theta,\varepsilon_r''} \\ \left.\dfrac{d|T_P|}{d(\varepsilon_r'')}\right|_{f,d,\theta,\varepsilon_r'} \end{cases} \tag{7.11}$$

Since an angular position of the sample under test is the critical issue in the measurements reported here, the S-curves have been calculated as a function of incident angle θ at $f = 330$ GHz with other parameters specified in Table 7.1. Figure 7.16 shows magnitude $|S|$-curves for the materials under test. We can see that the variation of an angular position demonstrates higher sensitivity for $Im\{\varepsilon_r\}$ than for $Re\{\varepsilon_r\}$. However, with increasing θ, losses associated with $Im\{\varepsilon_r\}$ become insensitive to the angular position of the sample, since the power transmittance tends to zero near grazing incidence.

Comparison of these data reveals the role of slab thickness, in that the sensitivity increases with increasing d for a low-loss material. In order to reduce the measurement error caused by the angular position inaccuracy, it is recommended to use higher-incident angles. However, foreshortening effects lead to limits, so that useful transmission data can be obtained only with incident angles not exceeding 50° [9].

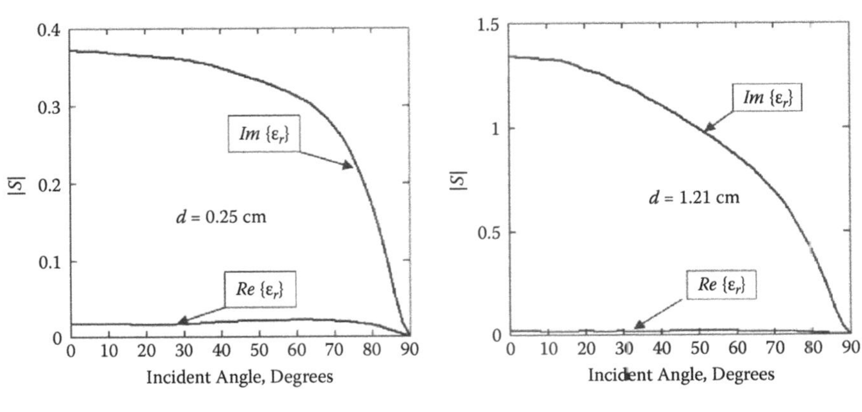

FIGURE 7.16
The calculated magnitudes of $|S|$-curves for the real and imaginary parts of the permittivity for thicknesses $d = 0.25$ cm and $d = 1.21$ cm.

There are several sources of inaccuracy in the measurement of complex dielectric constants:

- The error in an evaluation of the power transmission coefficient caused by a drift of the measured RF power
- The error in an evaluation of the frequency that may lead to changing calibration conditions
- The error associated with a surface roughness and thickness variations of the samples under test
- The error caused by an inaccuracy in determination of the incident angle

The most important factor is a drift of the measured RF power due to temperature instability of the absorbing thin-film element. Observation of this effect reveals that during a 30 min period such a drift is about 3% of the averaged value, resulting in 3% error in the measurement of the absolute RF power and power transmittance. Table 7.8 summarizes the results of error analysis caused by the above factor for the ceramic tile. The expected errors are 1.3% and 3% for real and imaginary parts of the dielectric constant, respectively. Taking into account the contribution of other factors, these errors can be evaluated at 2% and 5%, correspondently.

7.6 Characterization of Materials: Incoherent Illumination

7.6.1 Introductory Notes

In the previous section, characterization of common materials has been considered, assuming that the emitted radiation is coherent. However, passive millimetre and microwave sensors are based on the analysis of incoherent signals like the thermal emission of the human body in the mm-wave range. Since any practical sensor has limited operating bandwidth (BW), we can suppose that spectral density of the probing signal and the bulk complex permittivity of a material under test are constant within this BW. The latter is also valid for mm-wave imagers employing a frequency limited noise illumination [7, 36].

In the case of a heterogeneous medium such as granular materials with arbitrary shapes, some additional factors must also be taken into consideration: the shape of particles, frequency range, polarization effects, as well as coherency of the probing signal. However, theoretical and experimental studies for these materials are made difficult due to the ambiguity associated with the above-mentioned factors. The theoretical aspects have been well developed for homogeneous systems and for cases of particles of spherical and ellipsoidal shape. In many publications, coherent electromagnetic fields are employed in scattering characterization of granular materials [22], [23]. Theoretical predictions based on a quasi-static approach [24] are also widely used, limiting interpretation of scattering and depolarization effects in non-uniform materials when a characteristic dimension of irregularity is comparable with a wavelength. Such a situation is typical for the mm and THz ranges. In order to avoid polarization mismatching and coherent interference effects, an incoherent noise-like source can be useful as reported in [25], [26]. Intensive growth of ultra-wide band (UWB) systems needs a deep understanding of the mechanisms leading to clutter due to rough surfaces or non-homogeneous interfaces [27], which influence the spectral content of the response. Therefore, evaluation of scattering effects inherent to wide band (quasi-noise) signals is also becoming an important issue in the design of UWB sensors operating in complex, real-world environments.

The following description focuses on formulating the criteria for the application of both noise and coherent signals in microwave and mm-wave sensing and material characterization. In order to formulate such criteria, the double-layer dielectric structure is considered. It demonstrates typical oscillating patterns of transmission coefficient in the frequency domain which originate from the Fabry-Perot effect. To create a transmittance interferogram the adjustable air gap between the two layers is introduced. The setups, operating in the W-band (75–110 GHz), were assembled and tested in order to show the role of destructive factors in measurement microwave and mm-wave absorption using coherent and quasi-noise probing signals.

7.6.2 Modelling Interference in the Presence of Noise

As shown in Sections 6.5 and 6.6, multi-layer plane-parallel dielectric structures are good candidates for estimating the role of interference and noise averaging effects. Indeed, when the distance between layers is varied, a spatial distribution of the interference maxima (or minima) is observed. A periodicity, swing and height of interference maxima (or minima) depend on both real and imaginary parts of refractive index n and k (or complex dielectric constant $\varepsilon' = n^2 - k^2$, $\varepsilon'' = 2nk$). By recording the transmitted signal we can easily compare interferograms for coherent and noise illuminating fields and formulate a relevant noisy averaging criterion.

The configuration of a plane-parallel cell, employed in the experiments described below, is shown in Figure 7.17. It consists of two dielectric plates of thickness t separated by an air gap. Part of the transmitted energy is captured by a wide band detector. In order to create a transmittance interferogram, an adjustable air gap L is introduced.

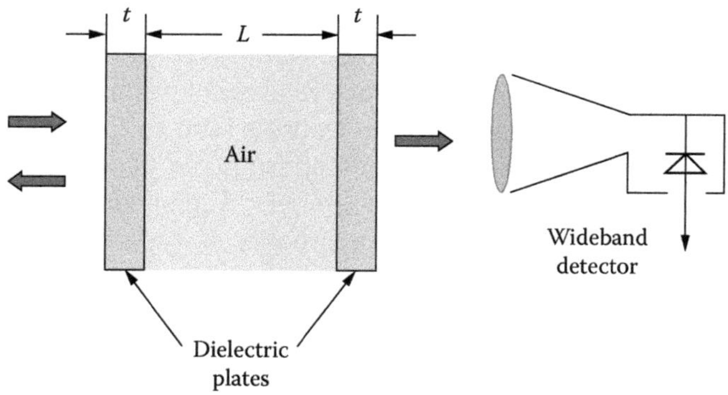

FIGURE 7.17
Configuration of the measuring cell used in mm-wave experiments. (From Kapilevich, B. et al., *Progress in Electromagnetics Research B*, 28, 35–54, 2011. Reproduced courtesy of The Electromagnetics Academy.)

Assuming that the measuring cell is illuminated by a plane wave, we can calculate the resulting transmittance matrix of the measuring cell T_c as a product of matrices describing separate layers [15,32], namely,

$$T_c = T_t \cdot T_L \cdot T_t \tag{7.12}$$

where

$$T_t = \begin{bmatrix} \cosh(\beta_t t) & \dfrac{i}{\sqrt{\varepsilon_{rt}}}\sinh(\beta_t t) \\ i\sqrt{\varepsilon_{rt}}\sinh(\beta_t t) & \cosh(\beta_t t) \end{bmatrix}$$

$$T_L = \begin{bmatrix} \cos(\beta_L L) & i\sin(\beta_L L) \\ i\sin(\beta_L L) & \cos(\beta_L L) \end{bmatrix} \tag{7.13}$$

$$\beta_t = 2\pi\sqrt{\varepsilon_{rt}}/\lambda,\ \varepsilon_{rt} = \varepsilon'_{rt} - i\,\varepsilon''_{rt},\ \beta_L = 2\pi/\lambda$$

Power transmittance T_p of the cell considered can be presented in the following form taking into account its symmetry [33]:

$$T_p = \left|\frac{2}{2T_{c(11)} + T_{c(12)} + T_{c(21)}}\right|^2 \tag{7.14}$$

where $T_{c(11)} = T_{c(22)}$; $T_{c(12)}$ and $T_{c(21)}$ are elements of transmittance matrix T_c.

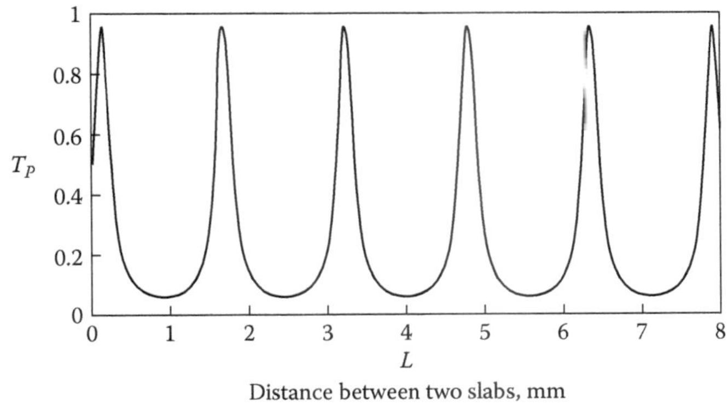

FIGURE 7.18
Interferogram of the power transmittance as a function of distance L. (From Kapilevich, B. et al., *Progress in Electromagnetics Research B*, 28, 35–54, 2011. Reproduced courtesy of The Electromagnetics Academy.)

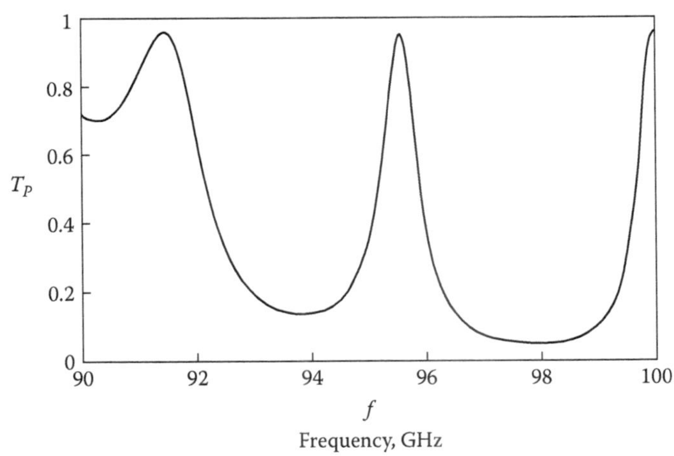

FIGURE 7.19
Interferogram of the power transmittance as a function of frequency f. (From Kapilevich, B. et al., *Progress in Electromagnetics Research B*, 28, 35–54, 2011. Reproduced courtesy of The Electromagnetics Academy.)

The typical behaviour of power transmittance as a function of distance L calculated using Equation 7.14 at a frequency of 96.5 GHz is depicted in Figure 7.18 for the Al_2O_3 sample with $t = 2.19$ mm, $\varepsilon'_{rt} = 9.6$ and $\tan\Delta = 0.001$. The behaviour of the power transmittance in frequency domain is shown in Figure 7.19 for the distance between plates $L = 30$ mm.

Interferograms depicted in Figures 7.18 and 7.19 correspond to the coherent illumination of the dielectric plates. When the probing signal is a frequency-limited noise we need to introduce some assumptions:

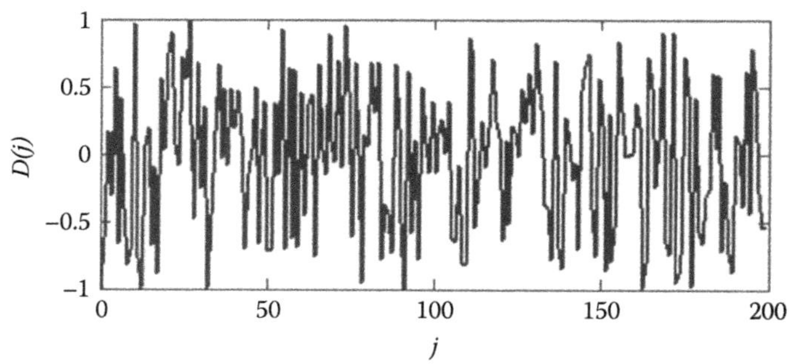

FIGURE 7.20
An example of normalized uniform random distribution of the 200 random frequency components. (From Kapilevich, B. et al., *Progress in Electromagnetics Research B*, 28, 35–54, 2011. Reproduced courtesy of The Electromagnetics Academy.)

1. Noise spectra occupy the band $f_c - BW/2$ to $f_c + BW/2$, where BW is the bandwidth of the frequency-limited noise signal.
2. Power spectral density is a uniform randomly distributed function within the specified BW.

Following [27], the total averaged power transmittance T_{av} can be written as

$$T_{av} = \int_{f_c - BW/2}^{f_c + BW/2} T_p(f) D(f) df \qquad (7.15)$$

where $D(f)$ is the normalized spectral density distribution.

As an example, the tested $D(j)$ function of the unit amplitude is depicted in Figure 7.20 for 200 frequency components randomly and uniformly distributed within the normalized bandwidth Δ%. The averaged transmitted power T_{av} of this distribution has been calculated from Equation 7.15, 'injecting' in carrier frequency f_c the frequency-limited noise of the bandwidth Δ varying in the range 2–100%. A random number generator has been used to imitate such an injection. The results of calculations are shown in Figure 7.21, where we can clearly observe degradation of the transmittance interferograms with increasing Δ. Beginning from the value of Δ = 30% and more, the noise-averaged power transmittance tends to its weighted median value near $T_{av} = 0.45$ for the situation discussed.

7.6.3 Experimental Verification of the Noise Averaging Effect

In order to verify the validity of the suggested model, a mm-wave setup similar to that described in [14], [15] has been assembled for operating in the W-band. It allows recording the frequency-limited incoherent transmittance interferograms using a wide band pyroelectric operating in the range

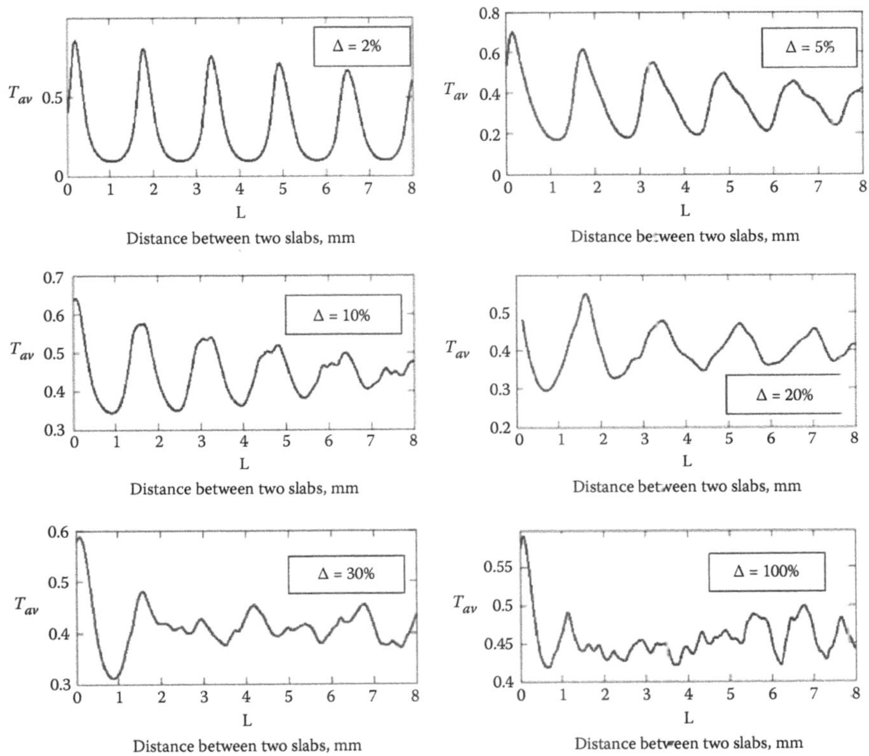

FIGURE 7.21
The calculated power transmittances corresponding to the random distribution shown in Figure 7.20 for different frequency limited noise bandwidth Δ varying in the range 2–100%. (From Kapilevich, B. et al., *Progress in Electromagnetics Research B*, 28, 35–54, 2011. Reproduced courtesy of The Electromagnetics Academy.)

0.03–3 GHz and a full W-band noise source. The same experiments can be performed for coherent illumination replacing the noise source by a continuous wave W-band source.

The schematic of the experimental setup is shown in Figure 7.22. It consists of a W-band coherent source, (1) Millitech (model GDM-10-3013M), and a full W-band (75–110 GHz) noise source, (2) Farran (model WG-NS-10), with ENR 13–15 dB. The output power of the coherent source is quite high, and a direct reading attenuator (3) is added to avoid overloading the pyroelectric detector (8). The output power of the noise source is too low to be directly measured by this detector. Therefore, a wide band low-noise amplifier (LNA) (4) is added at the output of the noise source. Both coherent and incoherent sources are connected alternately to the horn antenna (5), whose aperture is blocked by a rotating chopper (6) forming the AM modulation with frequency 10 Hz. The high-performance mid-range travel linear stage (the programmable micro-stepper), model ILS-100pp, Newport Co., is used for

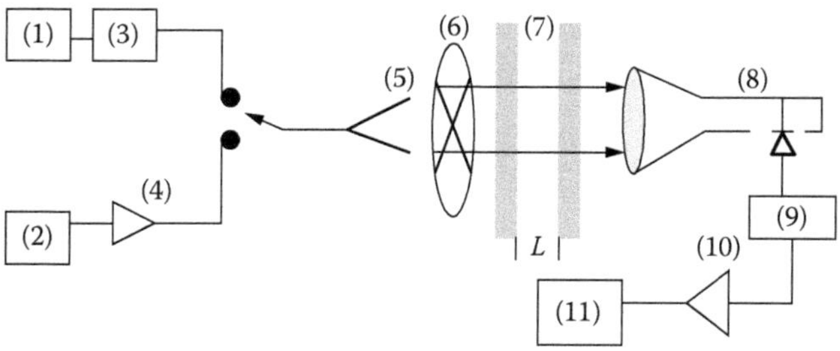

FIGURE 7.22
A schematic of the experimental setup operating in the W-band. (From Kapilevich, B. et al., *Progress in Electromagnetics Research B*, 28, 35–54, 2011. Reproduced courtesy of The Electromagnetics Academy.)

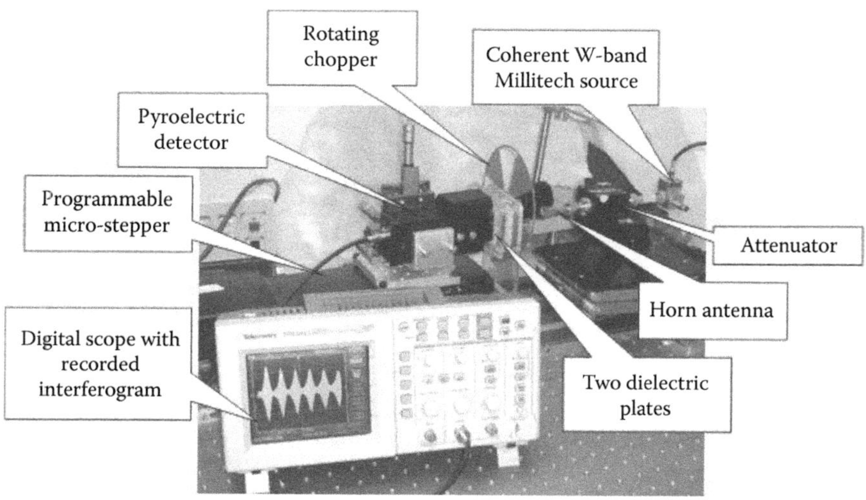

FIGURE 7.23 (See color insert.)
A view of the W-band experimental setup with a coherent source. (From Kapilevich, B. et al., *Progress in Electromagnetics Research B*, 28, 35–54, 2011. Reproduced courtesy of The Electromagnetics Academy.)

changing the distance L between the two dielectric plates (7) with an accuracy of about 2 μm. The slabs of thickness t = 2.19 mm were fabricated from Al_2O_3. One of the plates is fixed, while the second one is mounted directly on the input window of the detector so that both detector and the second plate are moving in the longitudinal direction simultaneously, varying the distance L between the dielectric plates. Transmitted energy comes to the input window of the MICROTECH pyroelectric detector. The detected signal is filtered by a BPF (9) with the central frequency f = 10 Hz (bandwidth = 1 Hz) and amplified by an operational amp (10). The Tektronix digital scope (11) is used for recording the detected signal. The general view of the W-band experimental setup is shown in Figure 7.23.

First, the coherent source was switched on at a frequency of 96.5 GHz and the transmittance interferogram recorded without the dielectric plates in order to estimate the beam divergence (Figure 7.24a). It is seen that the beam is rather uniform within the measured interval and the beam's widening effect is negligible. Then the two dielectric plates were installed and the coherent source was switched on. Figure 7.24b shows the measured transmittance interferogram as a function of the distance between the slabs for coherent illumination at frequency 96.5 GHz. Similar measurements were performed with incoherent illumination for the same plates as shown in Figure 7.24c. It can be seen that the variation of magnitude of transmitted signal corresponding to noise illumination is much less in comparison with the coherent mode, proving the concept of noise averaging.

The interferograms shown in Figure 7.24 were used for analysis using the standard median smoothing procedure applied to the absolute magnitude of the recorded signal. A finer structure of this signal's pattern can be reconstructed by employing a Gaussian kernel locally weighted averaging of the input data's vector vy' [29]. For each element in the n-element vector vy, the kernel smooth function returns a new vy'-element given by

$$vy_i' = \frac{\sum_{j=1}^{n} K\left(\frac{vx_i - vx_j}{b}\right) vy_j}{\sum_{j=1}^{n} K\left(\frac{vx_i - vx_j}{b}\right)} \tag{7.16}$$

where $K(t) = 1.078\exp(3.65t^2)$ and b is a bandwidth which is supplied to the kernel smooth function. The bandwidth is usually set to a few times the spacing between data points on the x-axis, depending on how wide a window is required when smoothing.

Finally, the results of such reconstructions after normalizing to incident power are depicted in Figure 7.25. We can clearly observe the quasi-periodical structure of the transmittance pattern inherent to coherent illumination as well as the essential averaging effect in the case of incoherent illumination. This effect leads to a reduction of the power transmittance variations in a vicinity of $T_{av} = 0.45$. That is quite close to the theoretical prediction obtained for the frequency-limited noise bandwidth $\Delta = 30\%$ shown in Figure 7.21. It should be pointed out that the bandwidth of the noise source used in experiments is about 25%, but the operating bandwidth of the pyroelectric detector is much wider. The input LNA limits the bandwidth of the transmitted channel. Actually, it is very difficult to estimate total transfer function of the transmitted channel, including both LNA and noise source. Hence, within the above assumptions we can state that a criterion of reasonable averaging is $\Delta > 25\%$ of total frequency-limited noise bandwidth.

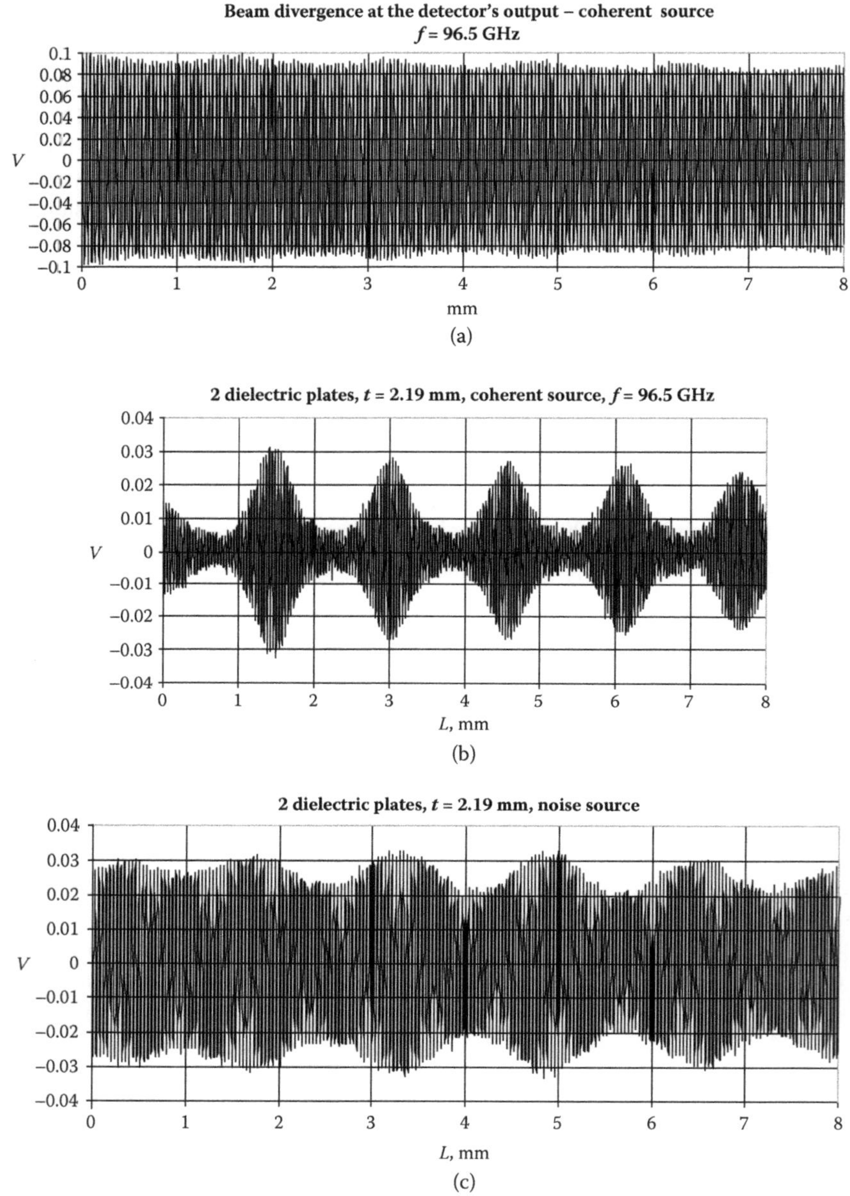

FIGURE 7.24
Illustration of the noise averaging effect: (a) transmittance interferogram without dielectric plates for coherent illumination; (b) transmittance interferogram with dielectric plates for coherent illumination; (c) transmittance interferogram with dielectric plates for incoherent illumination. (From Kapilevich, B. et al., *Progress in Electromagnetics Research B*, 28, 35–54, 2011. Reproduced courtesy of The Electromagnetics Academy.)

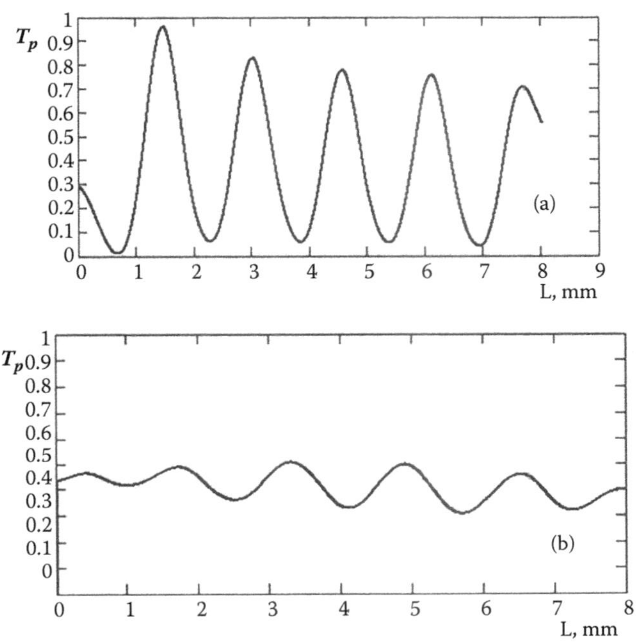

FIGURE 7.25
Results of reconstructing power transmittances for (a) coherent and (b) incoherent illuminations. (From Kapilevich, B. et al., *Progress in Electromagnetics Research B*, 28, 35–54, 2011. Reproduced courtesy of The Electromagnetics Academy.)

The effective attenuation of various materials has been measured using mm-wave pulse noise illumination setup shown in Figure 7.26a. The function generator is used for 1 KHz pulse noise modulation. The detected signal is depicted in Figure 7.26b. A calibrated W-band attenuator is connected with the input of the R_x module and is used for a direct reading of attenuation. The R_x antenna is beneath the gypsum plate.

Various building materials were tested. Some selected results of measurement of effective attenuation using coherent and incoherent illumination are presented in Table 7.9. Additional information about noisy coherence effects in material characterization can be found in [30]–[32].

All samples of materials under test were examined in orthogonal polarizations: horizontal and vertical. Most of building materials have a texture and demonstrate polarization sensitivity (anisotropy) of the measured attenuation. However, the gypsum plate is quite uniform (isotropic) and insensitive to the polarization of the irradiated wave.

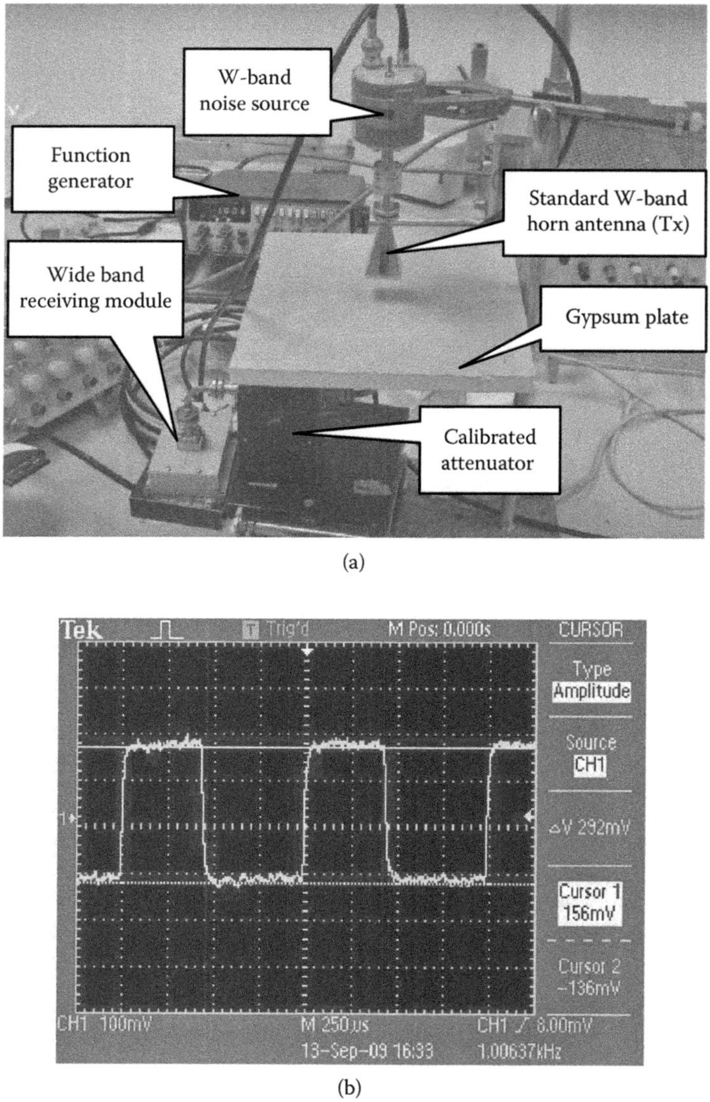

FIGURE 7.26
(a) The setup used in the W-band material characterization and (b) the screenshot of the detected signal.

TABLE 7.9

Measured Effective Attenuation in W-Band of the Selected Building Materials

Type of the Material	Effective Coherent Attenuation at 96 GHz (dB)	Effective Incoherent Attenuation (dB)	Polarization
One layer of wood board, 2 cm	–19	–14	Horizontal
One layer of wood board, 2 cm	–20.4	–15	Vertical
Corrugated carton, 0.5 cm	–3	–2.4	Vertical
Corrugated carton, 0.5 cm	–3.5	–2.2	Horizontal
One layer of gypsum board, 1.2 cm	–3.5	–3.2	Vertical or horizontal
One plate of cement tile, 2.5 cm	–39.3	–32	Vertical or horizontal
Plywood, 0.7 cm	–6	–4.5	Vertical or horizontal

7.7 Conclusion

The role of coating materials forming obscuring layers has been considered a factor responsible for degradation of a performance of the mm-wave sensors of concealed objects. Depending on the sensor's mode of operation, such materials should be characterized using coherent (active sensor) or frequency-limited noise (passive sensor) irradiated probing signal. The techniques relevant to mm-wave materials characterization were developed for both aforementioned modes. It was pointed out that noisy irradiation permits a better reach in suppressing undesirable interference between obscuring layers and essential averaging effect. The THz Matched Power Meter operating near Brewster's angle has been used for mm-wave material characterization in the range 100–330 GHz, allowing improvement of the measurement accuracy. The criterion of the angular-invariant conditions of measurement has been formulated to be applied to common obscuring materials such as fabrics, leathers, carton, etc.

References

1. M. S. Mirotznik, S. Yarlagadda, R. McCauley, and P. Pa, Broadband electromagnetic modeling of woven fabric composites, *IEEE Transactions on MTT*, 60(1), 158–169, 2012.
2. B. Kapilevich, Y. Pinhasi, B. Litvak, M. Anisimov, and D. Hardon, Complex permittivity measurements of textiles and leather in a free-space: an angular-invariant approach, *International Journal of Microwave Science and Technology*, 2012, article ID 375601 (2012). Available at http://www.hindawi.com/journals/ijmst/2012/375601/.
3. N. G. Paulter, E. N. Grossman, G. N. Stenbakken, B. C. Waltrip, S. Nolen, and C. D. Reintsema, Design of an active mm-wave concealed object imaging system, *Proceedings of SPIE*, 4373, 64–70, 2001.
4. L. F. Chen, C. K. Ong, C. P. Neo, V. V. Varadan, and V. K. Varadan, *Microwave electronics: measurement and material characterization*, Wiley, Hoboken, NJ, 2004.
5. W. D. Burnside and K. W. Burgener, High frequency scattering by a thin lossless dielectric slab, *IEEE Transactions on Antenna and Propagation*, 31, 104–110, 1983.
6. S. W. Harmer, N. Rezgui, N. Bowring, Z. Luklinska, and G. Ren, Determination of the complex permittivity of textiles and leather in the 14–40 GHz millimetre-wave band using a free-wave transmittance only method, *IET Microwaves Antennas and Propagation*, 2(6), 606–614, 2008.
7. B. Kapilevich, Y. Pinhasi, M. Anisimov, B. Litvak, and D. Hardon, Quasi-noise illumination in mm-wave imaging: from concept to realization, in *IEEE 27th Convention of Electrical and Electronics Engineers in Israel*, 1–4, 2012.
8. N. Gagnon, J. Shaker, L. Roy, A. Petosa, and P. Berini, Low-cost free-space measurement of dielectric constant at Ka band, *IEE Proceedings—Microwaves Antennas and Propagation*, 151(3), 271–276, 2004.
9. K. Meng Hock, Error correction for diffraction and multiple scattering in free-space microwave measurement of materials, *IEEE Transactions on MTT*, 54, 648–659, 2006.
10. B. Kapilevich, Y. Pinhasi, and B. Litvak, Measurement of complex permittivity of lossy materials in free space using matched THz power meter, *International Journal of Infrared, Millimeter, and Terahertz Waves*, 32, 1446–1456, 2011.
11. G. L. Friedsam and E. M. Biebl, A broadband free-space dielectric properties measurement system at millimeter wavelengths, *IEEE Transactions on MTT*, 46(2), 515–518, 1997.
12. K. Sato, T. Manabe, J. Polivka, T. Ihara, Y. Kasashima, and K. Yamaki, Measurement of the complex refractive index of concrete at 57.5 GHz, *IEEE Transactions on AP*, 44(1), 35–40, 1998.
13. F. I. Shimabukuro, S. Lazar, M. R. Chernik, and H. Dyson, A quasi-optical method for measuring the complex permittivity of materials, *IEEE Transactions on MTT*, 32, 659–665, 1984.
14. B. Kapilevich and B. Litvak, THz characterization of high dielectric constant materials using double-layer sample, *Microwave and Optical Technology Letters*, 49(6), 1388–1391, 2007.

15. B. Kapilevich, Y. Pinhasi, A. Yahalom, and B. Litvak, THz characterization of lossy materials using multi-layers measuring cell, in *33rd International Conference on Infrared, Millimeter, and Terahertz Waves*, California Institute of Technology, Pasadena, CA, 15–19 September 2008, paper 1130.
16. S. S. Rao, *Engineering optimization: theory and practice*, 4th ed., Wiley, Hoboken, NJ, 2009.
17. U. C. Hasar, A fast and accurate amplitude-only transmission-reflection method for complex permittivity determination of lossy materials, *IEEE Transactions on MTT*, 56(9), 2129–2135, 2008.
18. P. F. Goldsmith, Quasioptical systems, IEEE Press, 1998.
19. S. R. Saunders and A. Aragon-Zavala, *Antennas and propagation for wireless communication systems*, John Wiley, New York, 2007.
20. B. B. Mandelbrot, *The fractal geometry of nature*, W. H. Freeman, New York, 1983.
21. G. Franceschetti, A. Iodice, D. Riccio, and G. Ruello, Fractal surfaces and electromagnetic extended boundary conditions, *IEEE Transactions on Geoscience and Remote Sensing*, 40(5), 1018–1031, 2002.
22. G. Koh, Experimental study of electromagnetic wave propagation in dense random media, *Waves in Random Media*, 2, 39–49, 1991.
23. A. P. Toropainen, New method for measuring properties of nonhomogeneous materials by a two-polarization forward-scattering measurement, *IEEE Transactions on MTT*, 41(12), 2081–2086, 1993.
24. S. V. Gatash and L. D. Styopin, Dielectric permittivity of heterogeneous disperse systems, *Telecommunications and Radio Engineering*, 53(7–8), 200–203, 1999.
25. J. Polivka, Noise can be good, too, *Microwave Journal*, 66–78, 2004.
26. B. Kapilevich, B. Litvak, T. Ben Yehuda, and O. Shotman, Comparative characteristics of coherent and incoherent microwave scattering in dense inhomogeneous layers, *Russian Physics Journal*, 49(9), 913–916, 2006.
27. M. Ghavami, L. B. Michael, and R. Konho, *Ultra wideband signals and systems in communication engineering*, John Wiley & Sons, New York, 2004.
28. S. Cornbleet, *Microwave and geometrical optics*, Academic Press, Waltham, Massachusetts, 1994.
29. Mathcad Profeesional, Smoothing of X-Y data by kernel smoothing, data analysis documentation, 2001.
30. B. Kapilevich and J. Polivka, Noise versus coherency in mm-wave material characterization, in *33rd International Conference on Infrared, Millimeter, and Terahertz Waves*, California Institute of Technology, Pasadena, CA, 2008, paper 1139.
31. B. Kapilevich, M. Einat, A. Yahalom, M. Kanter, and B. Litvak, Millimeter waves sensing behind walls—feasibility study with FEL radiation, in *29-FEL—Proceedings of FEL 2007 Conference*, Novosibirsk, Russia, pp. 501–504.
32. B. Kapilevich and B. Litvak, Noise versus coherency in MM-wave and microwave scattering from nonhomogeneous materials, *Progress in Electromagnetics Research B*, 28, 35–54, 2011.
33. D. Pozar, *Microwave engineering*, John Wiley & Sons, New York, 2005.
34. S. Hallensleben, S. W. Harmer, and P. D. Townsend, Optical constants for the S20 photocathode and their application to increasing photomultiplier quantum efficiency, *Optics Communications*, 180, 89–102, 2000.

35. S. W. Harmer, P. Townsend, and N. Bowring, Enhancement of photomultiplier sensitivity with anti-reflective layers, *Journal of Physics D: Applied Physics*, 45, 055102, 2012.
36. B. Kapilevich, B. Litvak, M. Anisimov, D. Hardon, and Y. Pinhasi, W-band imaging detector with incoherent illumination, *Journal of Infrared, Millimeter, and Terahertz Waves*, 34, 9, 556–565, 2013.
37. B. Kapilevich, Y. Pinhasi, and B. Litvak, Application of the matched THz power meter for material characterization in free space, in *IEEE International Conference on Microwaves, Communications, Antennas, and Electronics Systems (COMCAS)*, Tel Aviv, Israel, 1–4, 2011.

Index

E

F

G

H

www.ingramcontent.com/pod-product-compliance
Ingram Content Group UK Ltd.
Pitfield, Milton Keynes, MK11 3LW, UK
UKHW022005180926
475701UK00003B/209

* 9 7 8 1 1 3 8 0 7 2 7 4 9 *